Major C.

Indian Notes About Dogs

Their diseases and treatment

Major C.

Indian Notes About Dogs
Their diseases and treatment

ISBN/EAN: 9783741109492

Manufactured in Europe, USA, Canada, Australia, Japa

Cover: Foto ©berggeist007 / pixelio.de

Manufactured and distributed by brebook publishing software
(www.brebook.com)

Major C.

Indian Notes About Dogs

.

INDIAN NOTES ABOUT DOGS

BY THE SAME AUTHOR.

Second Edition, Revised and considerably enlarged.   *Rs.* 2.

INDIAN HORSE NOTES,

An epitome of useful information arranged for ready reference on emergencies and specially adapted for Officers and
Mofussil Residents, all technical terms explained and simplest
remedies selected.

THACKER, SPINK AND CO., CALCUTTA.

INDIAN
NÓTES ABOUT DOGS,

THEIR

DISEASES AND TREATMENT.

COMPILED BY

MAJOR C ——— ,

AUTHOR OF " HORSE NOTES.

FIFTH EDITION

CALCUTTA

THACKER, SPINK AND CO.

1893

PREFACE TO SECOND EDITION.

It was· with a feeling of gratified surprise that I received within four months of the issue of the first edition, a letter from the publishers of " Indian Notes about Dogs" requesting the immediate preparation of a second edition.

I take the present opportunity of heartily thanking those gentlemen who have complied with the invitation contained in my first preface to write to me regarding any additions or improvements they could suggest. All such communications have received careful consideration, resulting in the addition of much new matter, especially regarding feeding and cleaning ; also regarding the treatment of mothers and their litters.

C.

.

PREFACE TO FIRST EDITION.

THE want of a moderately priced practical treatise on Dogs, especially adapted for India, has long been felt. This is proved by the applications made to publishers enquiring for such a work. There are, doubtless, many persons more qualified than myself to supply the demand; but as none of them have come forward, I thought it best to lead the way by printing some manuscript notes on the subject, which I have compiled for the use of myself and friends at various times during past years. A little work published by me at the close of 1878, styled "Horse Notes, by Major C.," especially adapted for officers and mofussil residents in India, met with a very flattering reception; but some correspondents have asked me whether I could not issue a fresh edition of it with a few chapters, treating in a similar, concise and practical manner, on Dogs. That has partly induced me to issue this work. No one can feel more than I do, that it is by no means perfect, and probably incomplete in some respects, though I have done my best with the time and means at my disposal to prevent these defects. However, such as it is, it is the first of its sort published in India ; and I shall feel greatly obliged to readers who can suggest improvements or point out defects and omissions, if they will write to me on the subject with a view to improvements in further editions. All communications to be addressed to " The Author of Indian Notes about Dogs," care of Messrs. Thacker, Spink and Co., Calcutta.

C.

CONTENTS.

Part I.

—

INTRODUCTORY REMARKS.

As a general rule, medicines have a similar effect on dogs as on human beings, with two great exceptions,—*viz.*, laudanum, of which a dog requires three times the dose for a man to produce the same effect; and calomel, which, the reverse of laudanum, produces extreme irritation of the lining of a dog's stomach and entrails, so that a man could take with impunity a dose that would kill two large dogs. As regards aloes, also, there is a peculiarity, for a moderate-sized dog could take as much aloes as would kill two men.

Carbolic acid is bad for dogs, and should not be used, as the skin absorbs it into the system, when it impedes the action of the heart, and may thereby cause death.

THE PULSE of a dog can be felt on the inside of the fore-leg, above the knee, or by placing the hand on the lower part of his chest. In health there are from 90 to 100 beats per minute.

PRACTICAL RULES.

To GIVE PILLS OR POWDERS TO ORDINARY DOGS.—Sit down on a chair, place the dog on his hind legs between your knees with his back towards you. Tie a towel round his shoulders to prevent his resisting with forelegs. Force open his mouth by pressure of forefinger and thumb on the *lips* of the upper jaw, which prevents his biting your fingers; with forefinger of other hand pass the pill as far back in the throat as possible, keeping nose well up in the air. Let the dog close his mouth, but don't let his nose down for some time to prevent his vomiting the medicine up, which he can easily do at will if his head be free.

Balls of nauseous ingredients should be

wrapped in thin paper and covered with sweet oil; or, if the dog is not off his feed, the pill may be hidden in a small bit of meat, which the dog will swallow of his own accord. Powders are usually mixed with equal quantities of sugar, and placed far back on the tongue, the mouth having been opened as here described. A powder may also be given by mixing it well with butter and smearing it, as if in fun, a little at a time on the nose, when the dog promptly licks it off.

To GIVE PILLS TO A BIG OR SAVAGE DOG. —Two persons required. Put the dog with his rump in a corner, straddle across him, put a strong cloth into his mouth, bring it together over the nose, where it is held firmly with one hand, and the other hand with another cloth similarly holds the lower jaw, whilst your assistant inserts the pill as above.

To GIVE DRENCH.—Hold him as for a pill, and pour the physic down his throat with a spoon or from a sodawater bottle. Only give

a little at a time and let the dog shut his
mouth, or he cannot swallow the dose. Then
repeat till all is finished, after which his head
must be kept up till you feel sure he will not
vomit the physic up.

ANOTHER MODE OF GIVING A DRENCH.—
Raise the dog's head gently and draw aside
the corner of the mouth so as to pull the loose
cheek from the teeth. Pour as much of the
liquid as can be done without spilling into
the pouch thus formed, from a spoon or
pewter squirt. The fluid will trickle down,
and the dog must swallow it, when more is
to be given, till gradually the whole dose is
consumed. This plan requires much patience,
but is useful when a dog will not open his
teeth except from sheer force.

TO GIVE ENEMAS TO A DOG.—Place him
on his side on a table, with one person to hold
him there, whilst another inserts the pipe (well
greased into the rectum, and pumps the fluid,
usually one to two pints) inside. The ordi-

nary clyster for human beings will do ; don't insert more than 2 or 3 inches of the pipe.

To MUZZLE A DOG FOR OPERATIONS.—If possible, get proper wire muzzle : if you can't, then use a cone of leather sufficiently large to enable him to put his tongue out, and well pierced with holes for fresh air, to be connected to the collar by a strap on either side.

For emergencies, bind a piece of strong tape round his jaws as near as you can to his eyes. Tie it in a knot between his eyes, and fasten the ends over his forehead to the collar to prevent the muzzle slipping off.

To BLEED.—Cut hair off a spot close to windpipe. Tie a string tightly round neck till you can feel the vein rise between the string and the head. Stick lancet well into the swollen vein, and bring it out in such a way as to make inside of the hole large enough for blood to escape. Instead of the neck, a vein on the inner side of a forearm may be used.

The quantity of blood depends so entirely on the size of the dog and its condition, that no rule can be laid down beyond avoiding fainting from taking too much blood. The maximum quantity would be about one ounce of blood for every three pounds of the animal's weight.

To STOP THE BLEEDING, remove the string and run a pin through the lips of the wound, and wrap some tow or thread round the ends of the pin (the point of which should be then cut off). After four days, remove the pin, leaving the tow or thread to fall off by itself.

To SETON.—Pick up a fold of skin and stick a lancet or knife through it. Through this hole pass a piece of tape smeared with blistering ointment, by means of a large bodkin or seton-needle, and tie the two ends of tape loosely together. Take care the dog does not bite or scratch the place, using muzzle, if necessary, for this purpose.

The object in using setons is to promote a

discharge of matter from any particular part and keep up an irritation there.

To VACCINATE.—Prick the thin skin on inside of ear obliquely, with a lancet, four or five times, the point of the lancet being freshly dipped each time in good vaccine lymph. This is believed by some persons to be an effectual guard against distemper, but is of doubtful efficacy.

To GIVE CHLOROFORM TO A DOG.—Make him lie down on a table, patting and soothing him till he is quiet. Fold a napkin into a conical shape, pour about twenty drops of chloroform into the small end, and hold the open end over the dog's nose and mouth so that he inhales the drug. Don't remove this till he is insensible; this is shown by the animal not wincing if the white of his eye is touched. If he shows any signs of returning sensibility before any required operation is finished, a further similar application of chloroform will be necessary.

ACCIDENTS AND INJURIES.

A small wound is best left alone, as the dog's tongue will keep it clean if accessible ; and if any healing ointment were applied, the dog would lick it off. If, however, the injury to the skin be on any spot, such as back of the head, where the dog cannot lick it, first wash the place clean with warm water, and when bleeding has stopped, apply common sticking-plaster. Don't use carbolic acid in any form, as it is injurious to dogs, though so good for horses.

If the wound be very large, the sides must be drawn together by one or two (as few as possible) plain stitches of strong silk. The dog will try to pull these out, so it may be necessary to muzzle him. Put a large bandage over the stiches, and change it often, or it will get foul. If the red granulations show above the skin, touch them lightly with damp bluestone (sulphate of copper) once a day. The dog's skin only heals by granulation, and not, as in human beings, by the torn sides of a wound re-uniting.

BROKEN BONES.

The fracture is usually plainly visible, but in any case can be felt by finger. If a rib is broken, at once tie a flannel bandage (or a horse girth will do for emergency) tightly round the chest: tie the dog to his bed to prevent his jumping about, and feed on slops; keep bowels open, and if restless from pain, give sedatives of opium, one grain being an average dose. These instructions apply equally in all cases of fracture. A broken leg should be set first by pulling the two ends apart, then tightly wrapped in flannel, and two wooden splints applied and kept on for three weeks or a month for a foreleg and six weeks for a hindleg.

DISLOCATIONS can be felt by hand. If of knee or toe, two persons should at once

reduce the dislocation by pulling strongly the
two parts of the limb injured in opposite
directions till the joint slips back into its
proper place; after which rest and low diet,
to prevent inflammation, are necessary. If
the hip, as is often the case, be dislocated,
professional assistance is indispensable; whilst
it is being sought, keep the dog quite quiet,
giving sedatives, if necessary, for that purpose.

SPECIAL TREATMENT OF BITCHES AND PUPS.

There is no fixed rule as to when bitches
will be in heat; usually it is every sixth
month, but it varies from every fourth to
every twelfth month. The first indication is
a slight enlargement of the teats for a week
or so, before the vagina commences to swell.
Heat lasts three weeks; after the first week
the bitch bleeds at the vagina for three or
four days, during which she should not be
crossed by a dog; but directly the bleeding

stops, say twelfth day from the commencement, is the best time for breeding purposes.
In the meanwhile she should either be provided with a pair of drawers or kept out of
the way of dogs. It is usual to allow the
selected sire to have two visits, though one
probably suffices. After this she must be
again isolated, or wear drawers, till the heat
ceases, or she will certainly mix the breed.

Steady but not violent exercise is beneficial
for a bitch in whelp, and she must not be
allowed to get too fat or too lean (that is, the
ribs should be clearly preceptible to the hand,
but not to the eye). Soup, milk and bread, and
oatmeal are the best food shortly previous to
confinement, which usually takes place sixty-
three or sixty-four days after copulation, and
is preceded by milk appearing in the teats,
and a mucous discharge from the vagina for
one or two days. On the appearance of these
symptoms a daily meal of cold-boiled sheep's
liver may be given, but no stronger aperient

is required. Don't put the bitch into a warm bath, nor check the vaginal discharge, which continues for a week after parturition.

The best arrangement for the confinement is a large board with a piece of old carpet nailed on it (to give foothold for the pups when sucking) and a raised edge to prevent the young ones tumbling off, with some straw on it for warmth and comfort.

Milk and water mixed and given several times a day, half warm, is the best food for a few days after confinement, when broth may be given and the ordinary food gradually resumed. A little daily exercise with liberal diet is conducive to the secretion of milk.

The tail and dew claws may be cropped before weaning, but ears should not be cut till the pup is four months old. The dew claws can best be removed with a pair of common wire-nippers. Weaning is usually done when pup is five weeks old.

An experienced person can tell during the

fourth week after copulation whether the bitch has conceived, and how many pups she will have, by placing the animal gently on her side on a table and manipulating the intestines. From the fourth to the seventh week, these signs are scarcely perceptible; but after that period any one can feel and count the pups. It is well to remember though that there is generally one more than can be felt by the hand.

Few bitches in India have strength to suckle more than three, or at the outside four, pups satisfactorily. The young have a trick of neglecting the two most forward teats because they are less comfortable than the others. In this case the fluid should be extracted from them by gentle squeezing between the finger and thumb. If this be neglected, the mother becomes restless, with hot tongue, dry nose, feeble appetite, and yelps when walking.

If all the pups die, it will be necessary to milk the mother by hand three times a day till secretion of milk ceases. An aperient should also be given, and the diet be limited. Rubbing the teats with salad oil and gin mixed in equal proportions often dries up the milk.

To bring up pups by hand get a baby's bottle with wash-leather teat. Prick the leather all over with a needle, and inside it place a small bit of sponge to afford resistance to the pup's tongue which naturally wraps round the teat in sucking. Each pup takes from ten to fifteen minutes for a meal, and requires 4 or 5 doses of cow's milk during the twenty-four hours. The sponge and teat should be frequently renewed as they get sour, though this may be lessened by keeping them in a solution of carbonate of soda. Weaning is effected by giving a saucer of milk, which they gradually learn to lap. At first, weaning puppies require to be fed four times daily, then three times, up to six months of age, and twice up to one year ; as a rough guide to quantity, allow one ounce per diem of food for every pound of the pup's weight.

" Worming " is an obsolete and useless barbarity which should never be revived.

RULES FOR SELECTING PUPS.

Fatness is desirable as a sign of good constitution. Every full-grown dog ought to have a black nose, but the color when first born is invariably red ; it is therefore necessary to wait for a fortnight after birth, when a small bluish-black mark becomes perceptible. If this be in the centre of the slit which divides the nostrils, the nose will ultimately be black; if, on any other part of the nose, it will only be partially black ; if there is no such mark up to three weeks of age, the nose will always be flesh-colored.

With Bull-dog and Bull-terrier pups, the slits formed by the closed eyelids should lie across the head and not be parallel to its length.

Greyhound pups should be held up by the toes of the forelegs, and those be selected whose hind quarters hang lowest.

If a pup of a sporting breed is held up by the tail, he should be able to put back his forepaws beyond his ears, as this is indicative of good shoulders.

RULES FOR FEEDING AND EXERCISE.

Feeding of all dogs should, so far as possible, be at regular hours, once or twice a day, but not oftener. The daily quantity of food should be from one-twentieth to one-twelfth of the weight of the animal, according to its work and exercise ; or one ounce of food for every pound of a dog's weight is a fair guide. For house dogs the best food is boiled oatmeal porridge, mixed with a little meat or broth to flavour it. Indian corn may be mixed in equal proportion with the oatmeal. The custom very prevalent in India of allow-ing dogs to habitually feed on remnants of curries and other highly spiced dishes from their master's table, is decidedly bad and conducive to diseases.

Dogs can fast for forty-eight consecutive hours and even longer without any injury; in this respect they differ diametrically from horses. Butter, fat, or grease as food render the skin diseased and make the body gross. Food ready cut up and served on a clean plate is calculated to cause excessive fat, costiveness and encrustation of tartar on the teeth; but an occasional bone, with a little dirt on it, will counteract this, and prove as beneficial to the canine species as the habitual use of a tooth-brush is for human beings.

A powerful dog in robust health can dispose of a meal of bones with great satisfaction to himself; but pet or house dogs not having their disgestive organs in such good condition, would suffer from the same meal. Nevertheless their instinct is, even when

pampered on cooked delicacies, to indulge sur-
reptitiously in gnawing some old bone devoid
of meat, which they often keep buried in the
ground for the express purpose. This instinct
should not be checked, but aided by supplying the animal with a large knuckle or other
bone, entirely denuded of meat. The larger
the better, as it is only for gnawing. The bones
of fowls, poultry or game should not, therefore, be given as they are easily crushed and
swallowed, when rugged portions of bone may
cause injury internally. The bone should be
thrown on the ground as a little earth will
improve it, taking care, of course, that it is not
on an ant's nest.

When oatmeal is not procurable or a change
of diet is desired, cold boiled rice with gravy
or broth poured over it as a relish, or dog biscuit may be substituted. For outdoor dogs
at hard work, paunch, tripe or liver are excellent meat; but the latter has a tendency, with
some dogs, to act as a mild aperient, and its

action must, therefore, be carefully watched at first. Beef and horseflesh are certainly bad for all except fox hounds or other large coarse dogs on strong work, both having a decided effect in producing tapeworms and skin disease. When meat, as above mentioned, is given, it should first be boiled, as raw flesh would be too stimulating for India.

It is advisable to watch a dog whilst feeding, and if it shows by slackening of the movements of the jaws and raising its head that it has satisfied its hunger, the unconsumed portion of the food should be removed to prevent over-feeding, with its usual results of deranged liver and stomach. Any ordinary dog-boy or mehter should have sufficient intelligence for this duty when once it has been explained to him.

If a dog declines to partake of its usual food, the better plan is not to offer any further refreshment till the regular hour for the next meal. But as loss of appetite is often the precursor of disease, the animal must be carefully watched for further symptoms. An occasional fast for a dog not at work is beneficial, and may ward off disease.

Potatoes, boiled thoroughly, mashed and served cold, are suitable for mixing with other food, or as a change. Cold boiled greens act as a gentle laxative, but few animals will eat them. Bits off a hot joint are decidedly injurious though evidently highly appreciated from the way badly trained dogs beg for them whilst their masters are at table.

Small house dogs may be fed entirely on dry crusts of bread, or biscuits soaked in milk or gravy, which should be poured boiling over the food and then allowed to cool ; a meatless bone being also given daily for gnawing.

It is a great mistake to suppose that anything unfit for human consumption will still be good enough for the dogs; any meat given to them should be fresh and wholesome, though it may be coarse. Nothing smoked or burnt, nor refuse, nor tainted flesh, is therefore suitable.

Dogs do not require salt to be mixed with their food. In fact it is a sort of poison for them, and a reference to the prescriptions contained in this work will show that it is recommended as an emetic. Mustard, pepper, and other table condiments, it need hardly be said, are still less appropriate.

For sick animals the following articles of diet are recommended, *viz.*, good *fresh* beef-tea, Liebig's extract, arrowroot, starch, powdered biscuit, ground rice, oatmeal and oswego. Any of these may be mixed with water, boiled to the consistency of common gruel and administered cold at intervals of an hour or half an hour as required. Whatever remains at night should be thrown away, and a fresh supply prepared daily, or in hot weather even twice a day.

To keep a dog in health, exercise is absolutely essential, and should be vigorous and exhilarating. If he can accompany his master in his morning ride or walk, it will be far more beneficial than the usual stroll with chain on under the care of the dog-boy or mehter. Exercise both morning and evening is desirable, but if only practicable once a day, let it be in the morning. Even when a master is unable to proceed further than his compound, he can still manage to give his dog some exercise by a chase after imaginary cats or rats, or fetching and carrying sticks, or even a good romp. During the long weary Indian day, with its compulsory confinement indoors, it is not a bad plan to teach dogs simple tricks as a mental diversion for them. Colonel Hutchinson's work called "Dog Breaking" gives good rules for instructing dogs.

In England animals frequently incur disease from defective sleeping accommodation. A damp or cold kennel will cause rheumatism, and even mere dirt will be prejudicial to health. In India, dogs usually sleep in their master's bedrooms as it is part of their duty to give notice of the approach of thieves or strangers. They should not be allowed to get on the bed, but should have a raised place of their own. Nothing can be better for this than an ordinary cane-bottomed dining-room chair, or a charpoy, which is also highly approved by the animal's during hot-weather; in the cold season it is only necessary to cover the seat with a blanket. If, however, a kennel be provided, it should be sufficiently raised from the ground, say, from one to two feet, to prevent damp by a free circulation of air underneath it. The straw should be changed daily, and the kennel occasionally washed with boiling water to prevent fleas and vermin accumulating. If, however, these pests do appear, treat dog and kennel as described here-after in Part III.

Instructions regarding the daily washing or cleaning will be found under the heading " Treatment in India of Imported Dogs," p. 93.

Part II.

WEIGHTS AND MEASURES.

1 grain	...gr.		
1 scruple	...	℈i =	20 grains.
1 drachm	...drm.(ʒi) =	60	„
1 ounce	...oz. (℥i) =	437·5	„
1 pound	...℔.	= 7000	„

MEASURES OF CAPACITY.

1 minim	min. (m)		
1 fluid drachm	fl. drm. (fʒi) =	60 minims.	
1 fluid ounce	fl. oz.(f℥i) =	8 fluid drachms.	
1 pint	... O	= 20 fluid ounces.	
1 gallon	... C	= 8 pints.	

For emergencies, when proper weights are not procurable, the following rough equivalents may be useful :

One rupee	... 3 drachms, or 180 grains.	
An eight-anna piece...	1½ „ „ 90 „	
A four-anna piece	... 45 grains.	
A two-anna piece	... 23 „ (nearly).	
A half-anna	... 200 „	
A pice	... 100 „	
Two Indian postage stamps ... 1 grain.		

For liquids, a common sherry glass usually contains two ounces, and a tumbler holds about 10 ounces or half a pint.

PRESCRIPTIONS FOR DOGS.*

(Numbered consecutively for facility of subsequent reference, and arranged alphabetically.)

ALTERATIVES (*i.e., which act very gradually on the constitution, and therefore require to be continued some time*).

1. A tablespoonful of cod-liver oil twice a day.

---

* Many of the ingredients mentioned are procurable in the bazaars, but as inferior drugs fail in the desired effect, and adulterated drugs are positively injurious, it is far better to get medicines direct from a respectable firm. I can most confidently recommend Messrs. Bathgate & Co., of Calcutta, who have supplied me for more than twenty years. I have never seen any article from them of an unsatisfactory quality. Their prices are reasonable, and they are extremely prompt in despatching anything required.—AUTHOR.

2. Or—

Ethiops mineral	...	½ ounce.
Cream of tartar	...	1 „
Nitre	...	2 drachms.

Divide into 16 or 20 doses, and give one night and morning, in all skin diseases.

3. Or—

Iodide of potassium	...	2 to 4 grains.
Liquid extract of sarsaparilla		1 drachm.

To be mixed and given in a little water twice a day.

ANODYNES (*to alleviate pain*).

4. For mild diarrhœa—

Castor oil	...	1 tablespoonful.
Laudanum	...	1 to 2 drachms.

Mixed as a drench, to be given every second or third day till diarrhœa stops.

5. For bad diarrhœa or purging—

Powdered catechu	...	1 drachm.
Prepared chalk	...	2 drachms.
Opium	...	6 grains.

Mix and divide into 12 powders ; give one every three hours in boiled flour or milk.

6. For spasms or cramp—

Laudanum	...	2 drachms.
Spirit of turpentine	...	1 drachm.
Sulphuric æther	...	1 ,,
Gruel ...	...	6 ounces.

To be mixed and injected as a clyster.

APERIENTS.

7. To act on liver—

Aloes	...	½ to 1 drachm.
Calomel	...	2 to 3 grains.
Oil of caraway	...	6 drops.

Mixed with little syrup to make a ball, and to be repeated every six hours till dog purges.

8. Or—

Calomel	...	3 to 5 grains.
Jalap	...	10 to 20 ,,

Mixed with little syrup to form a ball, is a strong purgative.

9. One pint of warm soapsuds as a clyster for mild constipation.

10.	Castor oil	...	½ ounce.
	Spirit of turpentine	...	2 to 3 drachms.
	Common salt	...	½ ounce.
	Gruel	...	6 to 8 ounces.

To be mixed and injected as a clyster.

Note.—Cold boiled liver is a gentle laxative. When castor oil is given alone, an ordinary dose is from two to four drachms. If castor oil is not available, olive oil in doses of two tablespoonsful can be used as a laxative.

BLISTERS.

11. Hog's lard ... 4 ounces. ⎫ Well mixed.
    Spanish flies ... 1 ounce. ⎭

12. For tumours, paint daily with tincture of iodine.

COUGH MIXTURE.

13.	Ipecacuanha wine	...	5 to 10 drops.
	Mucilage	...	2 drachms.
	Sweet spirits of nitre	...	20 to 30 drops.
	Paregoric	...	1 drachm.
	Camphor mixture	...	½ ounce.

Mix and give night and morning.

If cough is merely a symptom of worms, treat for *worms*.

CAUSTICS.

14. Nitrate of silver is the best. Bluestone (sulphate of copper) is milder and particularly good for the toes when ulcerated. Burnt alum in powder is a very mild caustic.

DIURETICS (*to cause increased secretion of urine*).

15. Ginger ... 2 grains.
Digitalis ... ½ grain.
Nitre ... 6 grains.

Mix with linseed meal and water and give as a ball.

EMBROCATIONS.

16. For rheumatism or strains—

Spirit of turpentine ... ½ ounce.
Liquor of ammonia ... ½ „
Laudanum ... ½ „

Mix, shake before using, and rub the mixture in.

17. Or—

Olive oil	... 6 ounces.
Aqua ammonia	... 2 ,,
Oil of turpentine	... 1 ounce.
Origanum	... 2 drachms.

Shake the bottle well, and rub the part with the mixture twice a day till it becomes hot and tender. Observe rest.

EMETICS.

18. One quarter grain of tartar emetic in solution, for middling-sized dog, to be repeated, if necessary, at intervals of two hours, will be found sufficiently strong for India.

Or a drench of one teaspoonful of salt and one of mustard in half a pint of tepid water.

Or a teaspoonful or two of common salt.

FEVER POWDERS.

19. Nitre in powder ... 3 to 5 grains.

  Tartar emetic   ... $\frac{1}{8}$ grain.

Mix and put dry on the dog's tongue every night and morning.

Or—

Sulphate of quinine	... 2 scruples.
Extract of gentian	... 1 drachm.

Mixed with sufficient syrup of ginger to make 20 pills, of which one should be given three times a day.

LOTIONS.

20. *For eyes.*—Sulphate of zinc, 20 grains, with half pint of water. Eyes to be washed night and morning with this, or one drachm extract of goulard with one ounce of distilled water.

21. *For penis.*—Mix one grain chloride of zinc with one ounce of distilled water.

22. *For canker in ear.*—Mix one scruple of extract of lead with au ounce of distilled water.

MANGE OINTMENT.

23.	Powdered aloes	... 2 drachms.
	White hellebore	... 4 „
	Sulphur	... 4 ounces.
	Lard or train oil	... 6 „
	Oil of thyme	... ½ ounce.

To be well rubbed on, and the dog muzzled to prevent his licking the ointment.

24. For red mange add one ounce of mercurial ointment.

RESTORATIVES (*after great fatigue*).

25. Cold tea with a little spirit in it.

26. Sherry and water.

TONICS.

27. Sulphate of quinine   ... 1 to 3 grains.

Extract of hemlock   ... 2   „

Ginger   ... 2   „

Mix and give twice a day.

28. Sulphate of zinc   ... 2 to 4 grains.

Extract of gentian   ... 3   „

Mix and form a bolus, to be taken three times a day.

WORM MIXTURE.

29. One scruple powdered glass mixed with butter as a ball, followed after six hours by two tablespoonsful of castor oil.

30. Or ½ drachm turpentine, 2 scruples areca (betul) nut. To be repeated three or four times, if necessary, at intervals of five or six days.

Followed by a dose of castor oil in four hours.

31. Or half drachm powdered areca (betul) nut, mixed with butter, in a ball to be given after the dog has been kept at least twelve hours without food ; to be followed after four hours by two or three tablespoonsful of castor-oil.

32. Or one to three drachms of " kamala " finely powdered.

WORM DRENCH.

33. Put from half to one ounce pomegranate bark in 1½ pint of water ; let it stand for twenty-four hours, then boil down to half the quantity and filter. Divide into three portions, and give at intervals of thirty minutes.

___

Part III.

EXTERNAL OR SUPERFICIAL DISEASES.

SURFEIT or BLOTCH, *from neglect or improper management, bad feeding, want of exercise.*

Symptoms.—Small eruptions on belly and thighs, roundish shape, about half-inch diameter : after few days scabs drop off, leaving spot red and bare. Not contagious.

Treatment.—Mild aperient (No. 7), regular exercise, vegetable diet, and good bedding, which should be frequently renewed.

MANGE, *from poverty, filth, contagion, or probably hereditary.*

Symptoms.—Skin red and cracked with offensive discharge and scabs. Hair comes off in large patches. Highly contagious. Parasites are present in the skin.

Treatment.—Same as above, but also apply the Mange ointment (No. 23).

Don't forget to muzzle the dog to prevent his licking the stuff.

A simple treatment, when the ointment (No. 23) cannot be readily procured, is as follows :—Five grains sulphur (*gundhuk* in Hindustani) three times daily ; also rub into the skin equal parts of lard and sulphur mixed. The skin should be well washed with yellow (not with carbolic) soap each time before applying this mixture, which should be done before a fire or in the sun to make the ointment flow freely over the skin.

RED MANGE, *from high feeding and basking before a fire as lap-dogs do.*

Symptoms.—Dog looks as if he had been sprinkled over with brick-dust. Commences usually on fore-legs, then on hind-legs, and after some weeks spreads to back.

Skin not scabby, and general health good, but constant scratching.

Treatment.—Treat as above. Apply the ointment (No. 24).

Don't use carbolic acid in any form for any skin disease of dogs.

INFLAMMATION, *from fleas, lice or ticks.*

Symptoms.—Constant scratching. By looking at roots of the hair the insects can be seen. Ticks have bloated bodies and spider-like legs, by which they hold on firmly. Lice often cover the body, and especially crowd upon the dog's head, around the eyes and the lips.

N.B.—Dog-lice will not live on a human being.

Treatment.—Keating's Persian insect powder is first-rate for temporary relief; when a fresh assortment of insects have assembled, use the powder again. If not available, try—

Acetic acid	...	4 ounces.
Borax	...	$\frac{1}{2}$ drachm.
Distilled water	...	5 ounces.

Well mixed and washed into roots of the hair.

Ticks should never be pulled, but cut off
with a pair of scissors. The Circassian insect
powder, sold by Messrs. Kemp & Co., Bombay,
is said to be very efficacious. If nothing else
be available, a little powdered camphor or
" butch" finely ground, rubbed on the coat, will
afford temporary relief from fleas. The kennel, if possible, should be changed; if not, it
must be thoroughly purified by pouring boiling water on it till all insects are dead, and
then painting the woodwork with spirits of
turpentine. Instead of straw give shavings
of yellow deal, if available, for bedding, or
substitute bran or sand for a short time.

Temporary riddance of fleas can also be
effected by rubbing a dog's coat well with
any soft soap and letting it so remain for half
an hour; then wash off the lather, and the
fleas, either dead or stupified, come away
with it.

Lice can be removed by saturating dog's
hair for twelve hours or more with common

castor-oil (about 2 or 3 ℔). This will act also
on the bowels mildly. The oil can be removed
by washing with yolk of eggs.

Maggots in sores may be got rid of by
placing freshly powdered camphor in the sore
or by a dressing of equal parts of spirits of
turpentine and oil, or by covering the place
twice daily with finely powdered *ghoor butch.*
This is procurable in most bazaars, and has a
decided curative effect on sores besides killing
the maggots.

WARTS, *from constitutional liability.*

Symptoms.—Red tumours inside the legs.

Treatment.—They should be removed by
the knife or ligature by a professional person.
Give the alterative (No. 2).

FOOT-SORE, *from exceptionally hard work,
such as a day's shooting after months of idle-
ness.*

Symptoms.—Besides weariness, that cuticle
covering the bottom of the dog's foot being
worn away, the sole is tender and raw.

Treatment.—Rest is required for nature to supply a new cuticle. Sponge the foot tenderly with tepid water to remove dirt, then dry with soft rag. Soak rags in a lotion of 2 grains of chloride of zinc, 2 drops essence of lemon, with one ounce of water, and bind them over the sore-foot, covering the whole with a bit of oilskin. Recommence work gradually when new cuticle is formed.

Bathing a dog's feet in tepid water in which a little powdered alum and common salt has been dissolved, will also afford relief.

INTERNAL DISEASES.

FEVER, *from exposure to cold or wet. Sometimes called* INFLUENZA.

Symptoms.—Loss of appetite, skin hot, shivering fits often at commencement, constipation, and highly-coloured urine. Often accompanied by running at nose and eyes, and cough as well, but no great loss of strength or condition.

Treatment.—Mild aperient (No. 7) ; no stimulating food, but vegetable diet. After aperient has ceased to act, give remedy No. 19; or if there be cough, give cough mixture (No. 13), followed by tonic if there be debility. Fresh air essential ; no violent exercise, though a walk may be allowed. Tonics must never be given whilst fever is actually on. For high fever, one-sixth of a grain of tartar emetic may be given every four hours till fever abates ; and if there be also constipation, as is common, a daily injection of two tablespoonsful of castor-oil in half a pint of lukewarm water will be beneficial.

DISTEMPER or TYPHUS FEVER, *from neglected fever or cold ; poison in the blood from infection.*

Symptoms.—Low fever with sudden and utter loss of strength and rapid emaciation, often accompanied by inflammation of the head or of some internal organ. Dung black and pitchy ; urine very high-coloured. Running from eyes and nose, and subsequently teeth also get covered with brown fur. Unless checked, the disease may prove fatal in a month.

Sometimes brain is affected, the signs of which are fits, or stupor, or delirium.

If bowels are inflamed, the dung will be black and very offensive, streaked with blood and shreds of coagulated lymph.

If skin breaks out in pustules filled with dark bloody matter, especially on belly and inside thigh, it is a favourable symptom.

Treatment.—Stop all solid food from first symptoms until complete recovery. No exercise, not even walking permitted, though a drive in fresh air is good. Thorough cleanliness of kennel indispensable, with plenty of clean, dry straw and fresh air.

The great point is to assist nature to get rid of the poison from the blood by the action of the bowels and kidneys. Commence with aperients (7 to 10) at intervals of two days.

Feed every two hours with teacupful of Liebig's Extract, warm, which, if not taken voluntarily, must be given as a drench to support the system. Port wine, mixed with arrowroot may be given occasionally, say every sixth hour, in lieu of the Leibig's Extract.

After the aperients have improved the look of the dung, mix antimonial powder, 2 to 4 grains, nitre, 5 to 15 grains, powdered ipecacuanha, 2 to 4 grains ; make into a ball and give twice a day. If there is much cough, add one grain of digitalis.

If the head be affected, seton the poll, as explained at page 6.

Directly favourable symptoms set in, give the tonic (No. 27).

If there be diarrhœa, give No. 4 or 5 according to severity of the diarrhœa.

RHEUMATIC FEVER, *from damp kennel or exposure to wet and cold, or from high feeding.*

Symptoms.—Low fever, with shivering; constipation; dog tries to keep in a corner and yelps with anguish if any one touches him or even approaches.

Treatment.—Hot bath and drying before a fire, followed by a good aperient (No. 8). Apply locally No. 16 or 17. Kennel must be dry and moderately warm.

For chronic rheumatism, give three times a week red herring mixed with a drachm of camphor to eat.

Loose flannel bandage may be tied round any joint affected.

MADNESS, HYDROPHOBIA, or RABIES, *occurring usually from three weeks to six months after a bite from a mad animal, or may arise from no assignable cause.*

Symptoms.—Great restlessness and complete change of temper first, a good-tempered dog becoming snappish even to his own master, snapping at the walls and at imaginary objects; great thirst and very impetuous style of drinking; indoors the animal persistently searches for places where other dogs have urinated, or where he thinks they have, and keeps licking the spot eagerly.

The dog in the earlier stage fights with his paws at the corner of his mouth to get rid of froth forming in his mouth. His gait is a long trot straight ahead without turning to either side.

Usually fatal in three or four days. Youatt states that the premonitory symptoms are obscure; first sullen, fidgetty, and constant shifting of posture; when curled up, the face is turned between the paws. Countenance becomes anxious and suspicious. A peculiar delirium, which causes him to snap and fly at imaginary enemies, is a certain sign of rabies.

If a dog not only continually scratches a sore ear, but rolls over like a football in so doing, be sure the sore ear was caused by the bite of a mad dog, and the poison is beginning to take effect.

Appetite becomes depraved, and animal eats his own dung or other filth. Sometimes organs of mastication become palsied, and food is dropped after being only partly chewed.

In the earlier stage, froth forms in the mouth, causing the dog to fight with his paws at the corner of his mouth as if a bone were sticking between his teeth. If this were the case, the mouth must remain permanently open, instead of occasionally closing. If after a time the dog tumbles over, be sure it is madness, and don't touch his mouth to look for a bone. Insatiable thirst soon supervenes.

If a mad dog makes any noise at all, it will be noticed that his voice is quite changed; his muzzle is always elevated, and a perfect bark is rapidly followed by a howl in a higher key.

Absence of sensation of pain is a characteristic sign. A mad dog will tear his own flesh or seize a red-hot poker with his teeth and make no cry if beaten.

In dumb madness the muscles of the lower
jaw are paralysed, the mouth is open, the
tongue blackened and protruding.

Treatment.—No cure.   Shoot the animal if
the symptoms are undoubted.   On the slightest suspicion of this disease, the dog should
be isolated from all living animals and closely
watched for further symptoms.

Note.—If dog bites human being or horse, apply lunar
caustic at once, and give aperient and cooling medicines
to ward off inflammation.

Tetanus and Lockjaw, *from severe injury.*

Symptoms.—If muscles of back be affected,
the body is contracted like a bow, till head
is close to the tail; if muscles of the belly be
attacked, the bow is the other way : sometimes
one side only is concerned.

Treatment.—Give chloroform to stop the
spasms, followed by purgative (No. 10), but
probably the disease will be incurable.

Turnside, *probably from a diseased brain
mostly attacks only highly-bred pups.*

Symptoms.—No frothing at the mouth, but animal turns continually round and round till exhausted.

Treatment.—Seton the poll, and feed on nourishing diet.

CATARACT, *from a blow, or constitutional.*

Symptoms.—A whiteness in the back part of the pupil of the eye.

Treatment.—Do nothing. Professional men only can operate for this.

AMAUROSIS, *from disease of the optic nerve.*

Symptoms.—Eye clear, no inflammation, but the pupil is much larger than usual, and the dog is blind of the eye affected.

Treatment.—Do nothing.

`INFLAMMATION OF EYE, *from distemper.*

Symptoms.—White of the eye turns bluish red, and is filmy; constant watering from eye, and dog avoids the light.

Treatment.—Tonic (No. 27 or 28) internally and good diet, will probably effect a cure.

The eye-lotion (No. 20) may be also used. Don't give aperients if there has previously been distemper, as the system is already low.

INFLAMMATION OF EYE, or common OPH-THALMIA, *when not preceded by other disease.*

Symptoms.—Eye inflamed, with thick discharge; white of the eye very bloodshot; dog avoids light.

Treatment.—Purgatives (Nos. 7 to 10), low diet, with warm fomentations (either milk or water) at first, followed by the eye-lotion (No. 20).

CANKER, or INFLAMMATION OF THE EAR, *from exposure or high feeding. Very common with long-eared dogs. Water-dogs get it from having the whole body, except ears, immersed in water, which naturally causes determination of blood to the ears, followed by inflammation.*

Symptoms.—Dog continually shakes his head and tries to rub or scratch the ear; the lining of the ear is red and inflamed, and the tips may probably be ulcerated. Dog sometimes becomes deaf.

If, in scratching, the dog rolls over like a football, he is mad from bite on ear by another mad dog. (*See* "Madness," page 43.)

Treatment.—Purgatives (No. 8 or 10) and low diet to reduce the inflammation. First bathe the ear with warm water and soft sponge. Then muzzle him, put his head flat and drop the lotion (No. 22) thoroughly inside the ear three or four times a day. The outside sores may be touched with wet bluestone. A water-dog should be prevented going into water whilst suffering from this, or he will get worse. A muslin bandage or cap may be advantageously tied round the head to prevent the ears being shaken about without causing too much heat.

Youatt gives the following rule for dropping lotion into a dog's ear. Two persons required. The surgeon must hold the muzzle of the dog with one hand and have the root of the ear in the hollow of the other and between the first finger and thumb. The assistant must then pour the liquid into the ear ; half a teaspoonful will usually be sufficient. The surgeon, without quitting the dog, will then close the ear and mould it gently until the liquid has insinuated itself as deeply as possible into the passages of the ear.

Note.—Take care of your clothes in this treatment, as the lotion containing lead will leave a white mark where it falls.

A remedy styled "Rackham's Ear Canker Specific" is sold in Calcutta, and has been favourably mentioned.

OZÆNA, or INFLAMMATION INSIDE THE NOSE.

Symptoms.—A stinking discharge from the nostril.

Treatment.—First syringe with warm water. and afterwards with a mixture of two grains of chloride of zinc with one ounce of water.

Keep bowels open by No. 7 or 9, and avoid heating food.

LARYNGITIS, or INFLAMMATION OF THE WINDPIPE.

Symptoms.—Hoarse and painful cough and quick, hard breathing with feverish symptoms.

Treatment.—Give active purge (No. 8) at once and seton the throat. Vegetable diet and fever medicine (No. 19) if symptoms do not abate.

PLEURISY, or INFLAMMATION OF COVERING OF LUNGS.

Symptoms.—Spasms of chest and shivering; laborious breathing, but air expired not hotter than usual; dry cough; pulse quick, small and wiry.

Treatment.—Don't blister. Give good aperient and only soup for food. Give pill three times a day of one grain calomel with one grain opium. Take care that dog is not exposed to cold or damp.

PNEUMONIA, or INFLAMMATION OF LUNGS.

Symptoms.—No spasms of chest; strong shivers; laborious breathing, and air expired is decidedly hotter than usual; pulse quick but soft; nostrils red; strong cough; crackling sound may be heard inside chest. Dog will not lie down much, but sits on haunches.

Treatment.—Same as above. If there is no pleurisy whatever, a blister may also be applied to the chest.

BRONCHITIS and ASTHMA.

Symptoms.—Shivering and continual hard cough, with discharge of mucus and wheezing; pulse full; no spasms, and cough apparently not painful. Bronchitis, if not cured, will result in asthma.

Treatment.—Don't bleed or blister. Give emetic (No. 18) followed by mild aperient

(No. 7 or 9) ; avoid stimulating food. If of long standing, give milk diet and beef tea, and rub chest with mustard liniment, or with No. 16 or 17.

CONSUMPTION.

Symptoms.—Very much as with human beings, *viz.*, cough, emaciation, blood-spitting, diarrhœa, and death.

Treatment.—Codliver oil (one to three teaspoonsful according to size of dog) three times a day to alleviate pain. Don't expect a cure.

INFLAMMATION OF STOMACH, or GASTRITIS, *from poison or improper food.*

Symptoms.—Vain efforts to vomit with much straining. Animal lies extended with belly touching the ground. Nose hot, quick breathing, great thirst.

Treatment.—Give emetic (No. 18) at once, followed by purge, and feed on slops only till all symptoms have disappeared ; then gradually resume ordinary feeding.

INFLAMED LIVER, or YELLOWS, *very common
with dogs exposed to wet and cold, or not sufficiently exercised and too much food.*

Symptoms.—Commences with shivers, hot
nose, feverish symptoms, clay-colored dung ;
white of eyes becomes yellow and vomiting
sets in, when dog soon gets exhausted and
dies. Liver, if enlarged, can be easily felt
by hand just below the right ribs.

Treatment.—Give promptly aperient (No. 7)
and a clyster (No. 9) after purging ceases.
Every four hours give pill of one grain calomel
and one grain of opium. Give exercise, and
rub No. 16 well over the right side. Only
slops for food till recovered. Take care the
dog does not get wet whilst under treatment.

INFLAMMATION OF BOWELS, *from bad feeding or result of colic.*

Symptoms.—Strong feverish symptoms, but
nose, ears, and legs cold ; evident pain on the
bowels being pressed by hand. Dog stands

with arched back, legs all together and tail pressed down.

Treatment.—Give hot bath and dry thoroughly; give castor-oil clysters and pill of one grain calomel and one grain opium every four hours. Don't rub the belly.

COLIC, *usually soon after improper meal.*

Symptoms.—Intermittent gripes, when dog howls with pain, with back arched and legs drawn together as in cases of bowel inflammation. Bowels *not* tender on pressure, and rubbing by hand affords relief.

Treatment.—Hot bath for thirty minutes, dry thoroughly; give castor-oil clysters, and rub the belly with No. 16. If pain continues, opium in 2-grain doses may be given every three hours.

DIARRHŒA, *usually from improper food or may be caused by internal inflammation.*

Symptoms.—If slight, only loose stools. If bad, the dung is slimy. If very bad, blood and white shreds or patches will be mixed with the dung, and dog becomes weak.

Treatment.—Attend to diet and probably no medicine will be required. If it be, give No. 4 or 5. If very bad, only rice-water to drink, and boiled rice and milk to eat.

CONSTIPATION, *from want of exercise or feeding on too stimulating food.*

Symptoms.—Belly hard and painful; what dung is expelled is very hard.

Treatment.—Give green food, such as porridge or oatmeal, and steady exercise; and, if necessary, the clyster (No. 9 or 10) or aperient (No. 7). Cold boiled liver is a gentle laxative.

PILES, *often accompanying constipation.*

Symptoms.—Dark nut-like knobs round the anus.

Treatment.—Soft food and exercise as above. A dose or two of castor-oil, followed by mix-

ing a little powdered brimstone, daily, with
the food till symptoms have gone.

INFLAMMATION OF KIDNEYS.

Symptoms.—Urine very scanty; great pain
in the loins.

Treatment.—Give twice a day a drink or
drench of five grains carbonate of soda and
thirty drops sweet spirits of nitre in wine-
glassful of water.

INFLAMMATION OF BLADDER *from catching cold.*

Symptoms.—Scaldings when passing urine
and discharge from penis of light yellow
matter (like gonorrhœa with human beings),
from which the exterior of the penis gets sore.

Treatment.—Give aperient, followed by
a dose of ten grains nitre with half ounce
epsom salt every third day. Keep penis
clean by frequent washing with warm water
(with soft sponge or syringe), and if it is
excoriated, apply No. 21.

STONE IN BLADDER, *possibly from drinking impure tank water, but cause can rarely
be correctly traced.*

Symptoms.—The animal constantly voids
small quantities of urine of an unhealthy
sort and, occasionally, drops of blood also.
The point of the penis remains protruding
from its sheath, and the leg is never raised
when urinating as in health. If compelled
to walk down steps or any steep declevity,
the animal shows the pain arising from the
stone, by sudden cries.

Treatment.—Only a professional man can
do permanent good in these cases. To alleviate pain give vegetable diet, and, if necessary, the anodyne No. 6.

BLOODY URINE (or Hæmaturia) *apart from
other symptoms of organic disease may be
seen.*

Treatment.—Cooling diet generally sufficient. If not give three minims tincture of
cantharides with two ounces of water.

PENIS INFLAMED, *from deranged digestive functions.*

Symptoms.—The dog constantly licks his penis, from which a thin fluid exudes. If not attended to, the discharge becomes thick and mattery, and the parts are tender and painful, and ultimately sores will form.

Treatment.—The food must be at once attended to, as that previously given was evidently too heating. Put the animal gently on his side or back, and first cleanse the parts thoroughly by squirting (or squeezing from a sponge) tepid water over the penis. If there be hairs matted together, cut them off. Then apply the lotion No. 21, three times daily, with a bit of lint. If that lotion is not procurable, either of the following, though not quite so good, may be used, *viz.*, half a scruple of alum with one ounce of rose water, or four grains sulphate of copper with one ounce of distilled water. If in cutting off hairs, the penis be accidentally cut, the bleeding can be stopped by applying powder of burnt alum, or a touch of lunar caustic.

PALSY *following attack of distemper or from worms, or from disease of brain and spinal chord.*

Symptoms.—Sometimes only twitching of head or one limb, but when bad, the whole body is affected, and also fits ensue frequently.

Treatment.—If dog has recently had distemper, give tonics (No. 27 or 28) and nourishing diet. If no distemper, suspect worms and give No. 29, 30 or 31, followed by tonics. Fresh air and steady exercise desirable.

FITS *from teething or from worms, or from brain disease.* (*See* " Palsy," " Apoplexy " and " Epilepsy.")

Symptoms.—A puppy lies on its side in convulsions, but no foaming at the mouth as in epilepsy.

Treatment.—If worms be suspected, treat accordingly; if a young pup be affected, try hot bath and attend to the bowels.

Sometimes a dog whilst out at exercise will suddenly stand still in a dazed manner, then with a loud strange guttural sound falls on its side, the limbs are ultimately rigid and violently contracted, the mouth is covered with foam, and both dung and urine may be involuntarily discharged. When the convulsions are over, the animal raises its head and stares about; if not prevented, it will rush off at a racing pace.

Treatment.—In such cases, wait quietly till the fit subsides, taking precautions, by slipping a rope or handkerchief through the collar or round the neck, to prevent the dog running away on recovery. It is useless to try any medicine whilst the fit is on, as the power of swallowing is suspended. If available, an enema of 2 drachms sulphuric æther, 2 scruples laudanum and 4 ounces of cold spring water may be at one injected and repeated after one hour's complete rest. As this class of fits is attributable to high feeding, the diet must be attended to, and the animal must be prevented from heating itself by running about violently for some days, or the fit will recur.

EPILEPSY, *from unknown case, probably hereditary.*

Symptoms.—Fit comes on quite suddenly and passes off nearly as quickly ; convulsion with foaming at the mouth, and blueness of lips and gums.

Treatment.—Bromide of potassium is the best remedy, two to three grains in a pill twice a day for a month.

APOPLEXY, *from too much blood to brain.*

Symptoms.—Heavy stertorous breathing with insensibility. Eye fixed and bloodshot ; no convulsions or foaming.

Treatment.—Bleed at once from neck vein. Give purge (No. 8) and clyster (No. 10). Keep bowels open afterwards, and avoid stimulating food.

WORMS, *from various causes, are very common in dogs' intestines.*—They are of three different kinds: the tapeworm, usually only one, of great length, flat and jointed; the round worms, two to eight inches long, like a common garden worm, except that it is pointed at both ends, and is of pinkish white colour; the maw worm, pointed at one end but blunt at the other, and of milky white colour, which only inhabits the larger and lower intestines.

Symptoms.—Depraved appetite and cough; staring coat, soon followed by loss of flesh; worm expelled with dung, which is evacuated frequently but in small quantities; breath offensive; nose hot and dry; sometimes fits occur.

(*See* " Fits.")

Treatment.—If symptoms are slight, give No. 29, 31 or 33, and repeat every five or six days till cured. If more powerful remedy be required, No. 30 will do. Change diet. Avoid "Indian pink," as it is a dangerous remedy. Five or six drops of spirits of turpentine in a dessertspoonful of castor-oil may be given to pups, and half a drachm similarly to a full-sized dog. Vermifuges should be given in the morning after a fast of at least 24 hours for ordinary dogs. With puppies, the last meal on preceding day should be omitted.

If you are certain there is a tapeworm, give one scruple powdered pomegranate root bark, followed in 4 hours by castor-oil (2 to 4 drachms). If there are only maw worms (which can be known by carefully inspecting the dung), they are best acted on by enemas, such as one tablespoonful of salt dissolved in a quarter pint of tepid water, or 30 grains aloes with a quarter pint of milk warmed.

WORM IN KIDNEYS.

Symptoms.—Bloody urine mixed with pus.

Treatment.—Treat as above for Worms.

DROPSY or ANASARCA, *from general debility or from kidneys not working properly.*

Symptoms.—Belly much enlarged ; emaciation; morbid thirst. Sometimes urine is mixed with blood.

Treatment.—If from debility, give tonics and nourishing diet. If urine be bloody, treat for kidney disease.

DYSPEPSIA, *from improper feeding and want of exercise.*

Symptoms.—Flatulence, loss of energy, alternate constipation and diarrhœa, extreme fatness, or else emaciated.

Treatment.—If too fat, limit the diet, give regular exercise, and occasional purgatives. If too lean, merely change diet every three or four days, and attend to general health.

C, ND 5

CANCER.

Symptoms.—Hard, knotty lump, which enlarges and ulcerates, and a red fungous growth appears. Most common about the private parts.

Treatment.—Incurable.

Note.—A cancer in the womb causes the vagina to appear as if the bitch were constantly in heat.

TUMOURS.

Symptoms.—Soft swelling just below the skin.

Treatment.—Must be cut out by some experienced person.

ABSCESS, *the result of inflammation.*

Symptoms.—Hard, painful swellings which gradually work their way to the skin and burst. Matter may be felt by its fluctuation when pressed by the finger.

Treatment.—First poultice, then stick a lancet in and cut the swelling open *downwards* to let all matter run out.

LIVE LEECH IN NOSTRIL.

Note.—Though this cannot be strictly called a disease, it often causes more discomfort and trouble than many ailments recognized by the Faculty.

Symptoms.—The dog constantly makes futile efforts by rubbing his paws over his nose and by snorting to get rid of something, and on careful inspection, the leech can be seen inside the nostril. This is very common with sporting dogs in Indian jungles.

Treatment.—Don't pull the leech out forcibly, or its fang will cause sore subsequently. First try injecting salt and water inside the nostril or rub half a cut onion over the nose, and seize the leech as it comes out of one nostril with the intention of going up the other. If this won't do, keep the dog as long as you can, from 12 to 24 hours without water, then hold its nose close to a saucer of water. As the leech comes down to drink, run a big needle into it to prevent its retiring, and then rub some common salt on it which causes it inevitably to release its hold,

Part IV.

DESCRIPTION OF VARIOUS KINDS OF DOGS.*

The Greyhound is of several kinds, *viz.*, English, Irish, Highland, and Lowland Scotch, Russian, Grecian, Turkish, Persian, and Italian.

All greyhounds hunt by sight rather than by scent; are not so highly gifted with sagacity as spaniels, and not generally so affectionate to their masters as other dogs are. A greyhound's age, like that of a race-horse, is reckoned from the 1st January of the year in which born.

The English Greyhound has peculiarly long attenuated head and face, terminating in singular sharpness of nose and length of muzzle.

---

* These descriptions are taken chiefly from Stonehenge, though considerably abridged by Youatt and others. It is only where they are silent, that the compiler has written original matter, as in the description of Japanese and Chinese dogs.

which gives a longer grasp, and therefore greater facility for securing its prey; ears should be close, sharp, and drooping; neck must be long to correspond with length of leg, or the animal is thrown out of his stride in trying to seize his game; chest must be capacious, but deep rather than broad; shoulders broad, deep, and obliquely placed; fore-legs set on square without bulging at the shoulder; legs straight with plenty of bone; the forearm (between elbow and knee) especially must be long, straight, and muscular; pasterns low placed, and toes neither turned in nor out. The back should be long and strong, though, if for use in hilly country, a dog with a shorter back would have the advantage; ribs arched; thighs and haunches muscular; hocks broad and low placed, or the animal will be devoid of speed.

Colour varies considerably, but for strength and endurance, the brindled, or brown or fawn coloured, is considered best.

The Scotch Lowland Greyhound is less speedy than the English, but stronger and larger, and has rougher coat.

The Highland Greyhound or *Deerhound* is like a large greyhound, but with a rough hairy coat, and in running he keeps his nose much higher than an ordinary greyhound would, so as to be ready to pull down his game. Height should be about 26 inches, and the dog may be of any colour. Unless possessing both speed to overtake game and courage to attack it, he is obviously not of much use. But if a cross with a bulldog be tried to improve the courage, the result would not be satisfactory, as the animals so bred attack a deer too much in front and thereby get thrusts from the horns.

The Irish Greyhound has shorter and finer hair than the Scotch one has, of pale fawn colour, and pendant ears. It is a large dog, sometimes four feet high, but not savage.

The Russian Greyhound is usually dark brown or iron grey, with soft thick hair, and the hair of its tail forming a spiral twist. It hunts by scent as well as sight, and is used in Russia for deer.

The Grecian Greyhound is smaller, and has coarser limbs than the English, and its muzzle is not so pointed.

The Turkish Greyhound is a small-sized hairless animal of little use.

The Persian Greyhound is very handsome, swift, courageous, but inclined to be ferocious. Though slighter than the English, he is not less enduring; his ears are pendulous and feathered.

The Italian Greyhound, very symmetrical but small, is more a pet than a sporting dog. It is bred on the coasts of Italy for sale to foreigners. It is usually good-tempered, though deficient in intelligence and personal attachment to its master.

Terriers.—Eight different kinds,—*viz.*—
(1) the English; (2) the Scotch; (3) the Dandie
Dinmont; (4) the Skye; (5) the Fox Terrier;
(6) the Bedlington; (7) the Halifax Blue Tan;
and (8) the Toy Terrier, of which again there
are various kinds.

The English Terrier usually weighs from 6
to 10 lbs.; it should be smooth-skinned, rich
black and tan, without a speck of white hair;
there must be a patch of tan over each eye;
nose and palate quite black; toes pencilled
with black more or less up the leg; nose long
and tapering; jaw slightly overhung; eye
small and bright, chest deep rather than wide;
shoulders powerful to let him dig away at
earth for hours without fatigue; loins short
and strong; fore and hind legs straight and
strong in muscle; feet round and hare-like;
tail fine with a down carriage.

The Scotch Terrier resembles the English in
all his good points, except that his coat is
wiry and rough and pepper-and-salt colour.

The real *Dandie Dinmont* is either reddish brown all over, or grey on back and light brown on legs, with silky hair on forehead, but rest hard and not long. Body long; legs short; ears large and hanging close to the head; eyes bright and full; tail straight and erect, with slight curve over the back. Weight should be 20 to 24 ℔s.

The Mongrel Dandie weighs just half the above weight, and has prick ears (*i.e.*, small standing up ears).

The Skye Terrier measures from tip to tail three times his own height. Ears may be either large, slightly raised, and falling over, or standing up like those of a fox. Tail must not curl over the back, and any white hair is a defect. A silky coat and jet black colour shows impurity of blood, though there are various kinds of skyes.

The Fox Terrier in olden days was attached to every pack of hounds to unearth the fox if necessary, but this has become obsolete from the quicker pace of all modern packs. The points of a fox terrier are, weight not over 16 ℔; colour white with black, or black and tan, or tan marks about the head; hair fine, but not silky; head flat and much wider between the ears than between the eyes; nose black; ears small, thin, and lying close to the cheek; light neck; full chest, but not deep; strong body and legs.

The Bedlington Terrier is a North country dog, has shortish body with long legs; colour liver or sandy, with cherry nose, or dark blue with black nose; head has, like the Dandie, a tuft of silky hair on top. Usually of very quarrelsome disposition.

The Halifax Blue Tan resembles the Scotch terrier, but coat is long and *silky*, blue on back and sides (any tan or fawn colour is a defect), with legs and muzzle a light yellow tan and beard several inches long of similar colour.

All terriers are very keen in pursuit of rats and vermin; but unless they have a touch of

the bulldog in their composition, they are
deficient in pluck if the vermin fights well.

———

The English Pointer.—If his pedigree can
be traced back to Mr. Edge's Kennel, it is
considered very satisfactory. The good points
are: head rather large and wide, with high
forehead; eye not too large; muzzle broad,
with outline square in front; neck long without ruffs (loose skin round neck). A distin-
guishing mark of pure breed is the tail,
which, though strong at the root, should suddenly diminish to within two inches of top,
when it becomes a fine point; shoulder-blade
must be long and strong; upper arm long,
with short forearm, with elbow well below
the chest, or he cannot stop and turn quickly,
and will soon get tired; foot must be round,
strong, with thick sole; coat should be short
and soft. White with dark head is colour
most preferred, as the white renders dog visible when working.

The Spanish Pointer is stronger and larger but less active than the English pointer.

The Portuguese Pointer is like the Spanish, only it has a bushy tail.

The Carriage (Great Danish or Dalmatian) *Dog* is spotted with black, or reddish brown, on white ground. The spots should be of uniform size and quite distinct from the white. The peculiarity of this animal is his love of horses, which induces him to follow where they go.

———

The Setter is so called because he used in olden days to drop or set down, instead of pointing, as he is now taught to do, to game. The Irish Setter is generally more hardy than the English one, but neither can stand heat without going into water at least every half hour. The good points which both sorts should have are: ears long and thin, and covered with soft silky hair, slightly waved; the tail (or "flag") must never curl over the back, nor, when in motion, should it ever be higher than its root; all four legs should be feathered (*i.e.*, have long curly hair), and the tail should have a fanlike brush of long hair.

The Irish Setter should be red with muzzle of same colour. A dark line down the back is objected to; the mouth should be black. The English setter has usually white ground with coloured (black, liver, red, &c.) head and patches, but there is no fixed rule.

Hair should be wavy and of silky texture throughout.

The Scotch Setter is like the English sort only his colour is black tan, or black tan and white.

The Russian Setter is little known in England, but would probably answer well in India, as he is not so easily knocked up by heat when working. He is covered with a thick matted coat, and as regards sagacity and nose is inferior to none, even if he be not, as some declare, superior to the British kinds.

Spaniels may be divided into three divisions :—

1st.—The Springer, which includes Sussex, Clumber, and Norfolk Spaniels.

2nd.—The Cocker, used principally for woodcock.

3rd.—Toy Spaniels, including King Charles, Blenheim, &c.

For sporting purposes, a spaniel under 12lb. weight is useless. The coat must be thick to stand consant wet, the nose must be first-class, or he will fail to follow game in concealed spots. The Clumber is invariably mute, but other spaniels will distinguish by their note what sort of game they are on. The tail should never rise above the level of the back, and its rapid working should show when the dog is on game.

The Clumber is a long, low, heavy dog, weight from 30 to 40 lbs., and height about 20 inches. Legs should be well feathered, and feet hairy ; coat thick, silky, and wavy, and colour of lemon and white.

The Sussex is about same height and weight as the Clumber, but shorter back, and has liver-coloured coat and nose. The great difference, however, is, that whilst the Clumber is generally mute, the Sussex has a full bell-like note.

The Norfolk Spaniel, which is most common is like a small setter; colour is usually black and white, or liver and white.

The Cocker is smaller and more active than the ordinary Spaniel, and the tail is usually cropped to prevent its knocking against bushes when working. It is used principally for woodcock.

Water-Spaniels may be recognized by toes being more webbed, and feet larger than with land dogs; coat woolly, matted, and oily to resist action of water on the skin. This oil makes the dog smell rather strongly.

The " M'Carthy " and " Doctor " breed of Irish South country water-spaniels are particularly prized. Colour must be entirely pure liver without any white; height 21 to 23 inches; ears 24 to 26 inches from point to point; well defined top-knot coming down in a peak on the forehead; tail short, round, stiff, without feather underneath.

Newfoundland Dogs are divided into two classes; the larger, 25 to 30 inches high, known as the large Labrador, is always mixed black and white; whilst the smaller, or St. John's breed, rarely exceeds 25 inches high, is usually quite black, though occasionally liver-coloured. In their native country they drag loads over snow. Though very companionable and gifted with acute power of scent they are no use for sporting purposes.

Bulldog (so named because formerly used for baiting bulls) should have round head; high skull; eyes not too large, with forehead well sunk between them; ears small, rather close together and not too far down; muzzle short with plenty of chop; back short and well arched towards the tail, which should be fine and of moderate length; coat rather fine; chest deep and broad, strong about the neck; legs muscular; foot narrow and well split up. The characteristics are great courage and extreme tenacity when once it has laid hold of anything, so that it is only by choking that it can be made to let go. A bulldog always attacks the head of an animal, and is not addicted to barking. Temper usually surly.

Mastiff (from the Venetian *mastino*, meaning large limbed) should have large head something between bulldog and bloodhound (but not showing front teeth as bulldogs do), with small drooping ear, small eye, deep voice; colour red or fawn, with black muzzle or various, but fawn and white is not considered good.

Poodles excel other dogs in intelligence and the ease with which they may be taught tricks. Though used in France for game, they don't seem to care about it.

Barbet is a sort of small poodle.

Maltese dogs should not exceed six pounds in weight; white colour, with occasional patch of fawn on ear or paw, resembling skye terrier, except that the coat is more soft and silky, and the tail curls over one hip.

King Charles' Spaniel should have black palate and nose; the latter very short and turned up, round head, prominent eyes, with a well marked 'stop' between them; long ears hanging close to the cheeks; colour should be black and tan, and weight as near 5 ℔., or even less, as possible; coat should be soft and wavy, but not curly; and legs must be well feathered down to the toes. Good watch dog generally.

The Blenheim Spaniel resembles the small King Charles, except in colour, which is white with red or yellow spots, and a white blaze between the eyes. The palate should be quite black.

Pugs should be fawn colour, with decidedly black face; coat short, silky and sweet smelling; feet like a hare's, and no dew claws on hind-legs; weight from 6 to 10 lb. Suitable for drawing-room pets only.

Japanese and Chinese Dogs.—Under this heading it is only intended to refer to the pet or toy dogs imported to England and India, as the majority of readers would take no interest in the 'chow-chow' and other varieties which are really to be found in ordinary life in China and Japan. The writer having travelled in Japan, has had the

opportunity of judging of the original articles in
its native country ; and he also served in the
China war of 1860, when the sacking of the
Emperor's Summer Palace caused the Chinese
pugs found therein to become fashionable
pets in Europe. Yet, with these facilities for
acquiring genuine specimens, he regrets having
nothing to say in favour of either Japanese or
Chinese dogs. They are, of course, utterly
useless for sport, or even killing vermin.

Doubtless, the fashionable taste for these
dogs will, like other fashions, die out, as they
are not to be compared to the genuine Maltese
for ladies' pets; but as it is certainly at present the fashion to profess to admire them,
some remarks regarding them are now given.
The 'points' have not been fixed in the pre-
cise manner of those of terrier and other English dogs. When the writer was in Japan, the
great test of excellence was for the purchaser
to endeavour to rest a silver dollar (about
double the size of a rupee) on its rim, on the

animal's nose. *If he could do so*, the price was half what it would otherwise be, as the prominent eye and extremely short nose of the perfect specimen rendered the attempt impossible. One peculiar test of the GENUINE Japanese, as distinguished from the Chinese of similar appearance, is its quasi-nautical mode of progression. If the latter wished to cross a room, it would go straight to the intended spot, whilst the Japanese indulges in a series of short curves right and left alternately. Another peculiarity of the genuine Japanese is its want of intelligence, ordinary sense, and affection for its owner; the Chinese possesses all the three attributes, though in a less degree than the King Charles' Spaniel, which may be considered the English improved descendant of the former.

Both Japanese and Chinese are prized in proportion to their diminutive size. The weight ought never to exceed five pounds. The skull should be round, with a very large round prominent eyes. If the colour be black and tan, the black must be very intense and rich, and the smallest touch of white anywhere would be a serious defect. The Japanese dog also has the appearance of incessantly weeping. There should be an indentation, or ' spot,' between the eyes. The lower jaw should project beyond the upper, and also turn up.

CROSSED BREEDS.

Retrievers are usually divided into two classes, called the curly-coated, and wavy-coated (or flat-coated). The former is covered all over, *except on* the head, with short, crisp curls of black or dark liver-colour without white. It resembles the small Newfoundland and the Irish Water-Spaniel or Setter, between which it is bred.

The wavy-coated retriever has the head of a setter, but shorter and less hairy ears, and the loose gait of the Newfoundland. Legs should not be much feathered; colour black; height between 20 and 24 inches. These dogs are expensive to keep, as they eat so much. Unless they can stoop well and have first-rate noses also they will not be able to find wounded birds.

Bull Terrier should be three-quarters terrier, or he will be too heavy and slow. He must have strong jaw, but the under one not projecting; strong chest; legs not bandy; tail fine and thin, like a bulldog, lightly and actively built height between 10 and 16 inches, and white colour is most prized, the best dog going for combining great pluck with sociable disposition and general intelligence.

The Lurcher is a cross between a greyhound and a sheep dog, used by poachers, as he combines speed and great hunting powers with silence.

The Cur is a cross between a sheep dog and terrier, and extremely useful in looking after his master's property, though a nuisance to other people from his habit of yelping at strange persons and animals.

Part V.

ON THE IMPORTATION OF DOGS TO INDIA, AND SORT TO SELECT.

MANY residents in India wish to import dogs from England, but have a great difficulty in deciding what sort to get. The descriptions of the various classes contained in this work will assist them in forming an opinion as to the correct name of the animal they would like to possess; but then remains the problem of its suitability to the climate. The climate of India varies so extremely according to locality, from the intense cold of the Himalayas, the equable temperature of the Cossyah, Garrow, and Naga ranges of hills, the humid moisture of Lower Bengal, down to the incessant steamy heat of the Western Coast, that it is impossible to lay down any rule applicable to the whole country.

The first point to consider is, what is the animal required for? It may be only for a lady's pet, or as a protector of the house from trespassers, or as a companion to its master in his solitary life in the interior of the country, or for sporting purposes. We will endeavour to give a few hints suitable for various cases. If required for a lady, the Maltese is, undoubtedly, most suitable, as its merry lively disposition, affectionate nature, and cleanly habits give it the pre-eminence amongst drawing-room pets. But if the lady also desires that her dog should be alert at night to detect, and by its barking to give warning of the approach of thieves (who can gain such easy access to Indian bungalows), then a very small terrier should be selected.

For companion to a gentleman we recommend the bull terrier. If resident in a hill climate, a Newfoundland is a very pleasant animal to possess, but he must be imported during the cold season, and on no account subject to the heat of the plains even for a week. For ordinary bird-shooting a "Cocker" Spaniel is probably about the best, or a Retriever.

MANAGEMENT ON BOARD-SHIP.

IF possible, an imported dog should, during the sea voyage, be under the charge of some friend, if the master himself be not on board, who will take the trouble every morning and evening to let the dog out of its kennel for a walk on the deck, and to see that the animal's habitation is not too close to the boilers, or unduly exposed to the sun's rays. The usual charge is five pounds for a dog's passage from England to India; besides which the butcher who feeds it will expect, and should receive, a tip of ten shillings at starting with the promise of a further sum of same amount if THE ANIMAL ARRIVES IN GOOD HEALTH at its destination. This gives him a direct interest in it, and he has the power of causing damage from neglect to any animal regarding which he has not received his "dues." But however willing he may be, neither he, nor any of the crew, can spare time to give the regular exercise which is conducive to health. If it be a water-dog, a bath twice a week (in fresh water, if possible, but if not procurable then in sea-water) is absolutely indispensable to keep the skin in order.

The expenses of importation being so heavy, it is obviously bad economy to bring any but really valuable animals.

The food on board-ship should be only two-thirds of the dog's ordinary allowance, whilst the quantity of fluid may be doubled. If, as frequently occurs, there are signs of consti-pation after the first few days of confinement on board, give a mild aperient, to be repeated as occasion requires.

Many animals, which are usually good-tem-pered, become snappish and surly on board, so

it must not be forgotten before starting to
supply not only the dog-chain, but also a
muzzle, for use whilst out exercising, or the
dog's daily walks on deck may be objected to
and prohibited. Plenty of clean loose straw
should be put in the kennel to lessen the
annoyance caused by the motion of the vessel.

TREATMENT IN INDIA OF IMPORTED DOGS.

IT is necessary to remember that a dog
recently from Europe requires comforts and precautions against the climate, just as much as,
if not more than, a human being under similar
circumstances. The latter will believe from
the information of others, without waiting to
ascertain the fact from painful personal experience, that undue exposure to the sun's rays
induces sickness, pain, and death. But the
poor dog, accustomed to move about freely in
the open air of his native land, cannot realize

this fact, and hence the necessity for keeping him shut up during the heat of the day. But if a punkah be necessary for his master, it is equally so for the dog, and water cooled with ice (though not with the ice actually in it) is a good tonic. The actual ice may injure the digestive organs. If the house is thoroughly closed, so that the dog cannot get outside, where any passing cur would be a sufficient powerful inducement to tempt him out under the sun's rays, the best plan is to let the dog roam about the house. But if there be any chance of his getting out of doors, he should be chained up in the same rooms where his master sits, with a plentiful supply of drinking water within reach, and both morning and evening he should have regular exercise. A water-dog must have a tub into which he can plunge when so inclined, or his skin will soon become diseased. For other sorts, a bath once a day is sufficient with thorough drying, and *to be followed* by exercise to avert the chance of rheumatism from insufficient drying. Warm-baths are debilitating, but the chill may be taken off the water if very cold.

Washing is not indispensable to cleanliness, and if the animal cannot be exercised after its bath, or always shivers or exhibits a marked antipathy to its bath, the following process may be advantageously substituted. Smear the yolk of hen's eggs, from which the white has been carefully removed, well into the dog's hair. For an ordinary sized terrier three eggs will be sufficient. Pour a little water on the animal's back and rub by hand briskly till the body is covered with lather, which can then be removed by pouring water over it. This need not be repeated oftener than twice a week, hand-rubbing, combing and brushing only being requisite on the intermediate mornings.

The aversion to a bath shown by some dogs may be traced to the fact of having

their heads forcibly immersed when the soapy
water causes their eyes to smart. If soap,
instead of yolk of egg, be used, it should be
of some very mild quality, *and clean* water
should be poured over the dog's head instead
of dipping it into the bath.

If a dog be much troubled with fleas, a
teaspoonful of spirits of turpentine should be
mixed with the yolk of each egg used for
cleaning him, and at the same time his kennel
must be thoroughly purified. This can be
done by throwing buckets full of boiling water
repeatedly over it to kill the fleas, and subsequently painting the woodwork with spirits
of turpentine. In England it is also recommended to give the animal the shavings of
yellow deal for its bed; but this might be difficult to procure in India. There are plenty
of nostrums which will instantly kill fleas;
but the difficulty is to select those which will
not injure the dog at the same time. The
Circassian insect powder sold by Messrs.

Kemp & Co., of Bombay, in pots from one rupee upwards, is very highly spoken of in that presidency, but the author has had no opportunity of testing it personally. For further information on this subject, reference should be made to Part III of this work under the heading of External or Superficial Diseases.

In England, house-dogs incur disease from basking too much near the fire (*vide* under the heading of Mange in list of diseases).

In India, the reverse process may occur from lying too close to thermantidotes and "khuskhus" tatties, which will perhaps induce rheumatism, sorethroat, cold, or other symptoms of illness. But this applies equally to human beings, so the great rule to observe is—

"WHATEVER PRECAUTIONS BE REQUISITE TO KEEP A EUROPEAN IN HEALTH ARE EQUALLY DESIRABLE FOR EUROPEAN DOGS."

From this it follows logically that it is

highly conducive to health to send them to the hills for the hot weather, and if the master goes there himself, or send his wife and family, he should certainly allow his English dog to accompany the party. But this may not be always practicable, and then strict attention must be paid to the foregoing suggestions.

Part VI.

HINDUSTANI VOCABULARY REGARDING DOGS.

NOTE.—The spelling is according to the ordinary pronunciation of English letters without any attempt to follow scientific systems. As local dialects vary considerably, and Indian attendants on dogs are illiterate, elaborate collections of vernacular synonyms are of little practical use. The following common terms may, however, be convenient for persons imperfectly acquainted with Hindustani, and save the trouble of reference to dictionaries on emergencies.

Aloes	... Moosubbur; elwar.
Anus	... Ganr.
Aperient	... Pait kholney ke dewai.
Assafœtida	... Heeng.
Astringent medicine	... Pait bund kurney ke dewai.
,, ointment	... Zakhm bund kurney ke murhum.
Back (of an animal)	... Peeth.
Bandage	... Puttee.
Bark, to	... Bhaunkbna.

Bathe, to (to give a bath)...	Goosul dena.
„ (to lave)	... Dhona ; dho dalna.
Belly	... Pait.
Birth or confinement	... Bachcha dena.
Birth (of pup)	... Junum ; paidarish.
Bitch	... Kothi ; kootiya. [larna.
Bite, to	... Dant se katna ; moouh cha-
Bladder	... Phookna
Blind of one eye	... Karnar.
„ both eyes	... Andha.
Blood (the fluid)	... Lohu ; khoon.
„ (descent)	... Zart.
Bluestone (sulphate of copper)	... Neela tootiya.
Body	... Budun.
Bone	... Huddee.
Bowels	... Untree.
Brain	... Mughz.
Breath	... Dum ; nufus.
Breathe, to (ordinarily)	... Dum lena.
„ (gaspingly)	... Harna pharna kurna.
Brimstone	... Gundhuk.
Bugs	... Khutmul.
Camphor	... Karfoor,
Canker	... Fora.
Caustics	... Tez dewai.
Carbolic	... Carbolic ke tail.

Chain	... Zunjeer.
Cheek	... Garl.
Chest	... Charteo.
Chiretta	... Chiretta.
Claws	... Narkhoon.
Clyster	... Pichkaree.
Cold or catarrh	... Surdi lugga.
Cold season	... Jharey ke mausum.
Cold	... Thunda.
Colic	... Pait men marora lugga ; koolinj.
Corpse of a dog	... Kootta ka larsh.
Costive	... Pait bund hai.
Cough	... Khanse.
Cry, to	... Rona.
Deaf	... Buhra ; khan se nay soonta.
Debility	... Cumzoree ; lartarkut.
Diarrhœa	... Dust lugga ; pait chulta.
Diseases of bladder	... Phookney ke beemaree.
„ bowels	... Untree ke ditto.
„ kidneys	... Gurda ke ditto.
„ liver	... Kuleeja ke ditto.
„ lungs	... Phepra ke ditto.
„ stomach	... Pait ke ditto.
„ skin	... Chumra ke ditto.
Drenching horn	... Deewai ke sing.
Dung of cow	... Gobur.

Dung of dog	... Kootta ke mila.
„ horse	... Leed.
Ear	... Karn.
„ canker in	... Karn men fora lugga.
Enteritis, or inflammation	
of bowels	... Untree men durd.
Eye	... Ankh.
Eyelid	... Poolook.
Female	... Mardeen ; marda.
Fever	... Bookhar ; tup.
Flea	... Pissoo.
Flesh	... Gosht.
Foam	... Kuf ; phen.
Foot	... Panw.
Fore-leg	... Hart, or ugla hart.
Ginger, dry	... Sont.
„ green	... Udruk.
Girth (for bandage)	... Zeen ka tung.
Granulate, to (as wounds),	Angoor hojaua.
Growl, to	... Bhoonbhun kurna.
Guts	... Untree.
Hair, short, on body	... Roan.
„ long, on tail or ears,	Barl.
Haunch	... Koola.
Head	... Sir ; khopree.
Heal, to	... Sookhna.
Heat (bitches), to be in	... Gum hona ; mustee.

Heat of weather	... Gurmee.
Hind-leg	... Pichley pair.
Hip	... Koola.
Howl, to	... Bhaunkna nkarna.
Howl, a	... Pukar; wawaila.
Howling	... Kootha ya bhariya ka rona.
Illness, severe	... Sakht beemaree.
,, slight	... Mandugee; beyaram hona.
Indigestion	.. Budhuzme.
Inflammation	... Sozish.
,, of eyes	... Ankh men larlee.
Jaw	.. Jubra.
Kennel	... Kootha ka ghur; sugkarna.
Kerosine oil	... Muttee ka tail.
Knee	... Ghootuar; zanoo.
Lave, to (or bathe a place)	... Dhona; dho dalna.
Lame	.. Lungra.
Lice	... Joon; chillur.
Lick, to	... Chartna.
Leech	... Jonk.
Lip	... Lub; honth.
Liver	... Kuleeja.
Lungs	... Phepra.
Mad	... Deewana; pugla.
Maggots	... Keera.
Male	... Nur.

Mange	... Koojlee; karrish.
Matter (from sores)	... Peeb ; rardh.
Mouth	... Mookh.
„ corner of	... Buchh.
Mustard	... Rai.
Muzzle for dogs	... Moonh bund kurnakachumra
Nail (of foot)	... Narkhoon.
Nerve	... Resha.
Nipple	... Thun ; choonchee.
Nitre	... Shora.
Nose	... Nark.
Nostril	... Nuthna.
Ointment	... Murhum.
Pain	... Durd.
Palate of mouth	... Tarloo.
Pant, to	... Harmphna.
Paw	... Punja.
Penis	... Laurar; hathiyar; dnnda.
Pregnant	... Garboen ; pait men bacha.
Pulse	... Nubz.
Pup, to	... Bacha dena.
„ a	... Pillar ; bacha.
Purgative	... Pait kholney ka dewai.
Rag, a clean	... Sarda chitra.
Rectum	... Gaur kee beetur.
Rib	... Parslee.
Rheumatism, acute	... Bart ; baee.

Rheumatism, chronic	... Agey ke bart; buhot roz ke bart.
Rump	... Pootha.
Saliva	... Rarl.
Salt	... Nimuc; noon.
Scrotum	... Andkos.
Shivering	... Kapkapi; karmpta.
Shoulder	... Phur.
Sinew	... Nus.
Skin	... Chumra.
Sleep	... Neend.
Sleepless	... Aneenda.
Snap, to	... Lapakna.
Snarl, to	... Garajna.
Sole of foot	... Talua; pubbar.
Sore, a	... Ghao; zakhm.
Sorethroat	... Kunth men durd.
Spine	... Reerb.
Sprain	... Moch; pechish.
Stomach	... Pait.
Straw	... Beecharlee.
„ a single	... Tinker.
Sulphate of copper	... Neela tootiya.
„ zinc	... Soofed tootiya.
Sulphur	... Gundhuk.
Swelling, a	... Phulao; sooj giya.
Tail	... Doom.

Teat	... Thun ; choonchee.
Teeth	... Darnt.
Testicles	... Koosiya.
Thigh	... Ran.
Throat	... Kunth ; gulla.
Tongue	... Jeeb.
Treacle	... Tiriyak ; rarb.
Turpentine	... Tarpeen ka tail.
Udder	... Lewa.
Urine	... Pishab.
Vagina	... Choot.
Whine, to	... Rona.
Womb	... Kokh ; dharan.
Worms	... Kenchwa ; keera.
Yelp, to	... Kutkutana ; cheeoheeyana.

INDEX.

OPINIONS OF THE PRESS.

Pioneer.—Major C., already favourably known as the author of a little work called *Horse Notes*, has just published a most useful and practical treatise, entitled *Indian Notes about Dogs, their Diseases and Treatment.*

Englishman.—* * * It gives in plain language a concise description of most of the ailments which dogs are liable to, either from disease or accidents, with directions for their treatment. The prescriptions are simple, and there is a Hindustani Vocabulary added, to enable any one to ask for the medicines required in the Bazar. * * * The Major has produced a very useful little book, which all those who possess dogs in India and care for their welfare will do well to possess.

Asian.—* * * We think we have said nearly enough to show the utility of the book so far as the treatment of diseases goes, but it contains a good deal of useful practical advice in addition. * * * There is also a useful chapter on the management of dogs on board-ship. * * * * Finally, the book ends with a Vocabulary of Hindustani words regarding dogs, and with this we think we have said quite enough to recommend it.

Civil and Military Gazette.—* * * We can, however, thoroughly recommend Major C.'s little book to all in this country fond of canine pets, as it contains much useful information in a small compass, and is not too learned or technical for the ordinary reader.

Critical Notices.

The Madras Times.—A very compact and handy little work. * * * Major C., an author already known to Anglo-Indians by his little work "Horse Notes," has placed before his readers a good deal of useful information, in very small compass, clearly expressed. * * * We can honestly recommend the few sentences on treatment in India of imported dogs to all those who desire to see their pets in good condition, sweet-scented, in capital spirits, and full of frolic and fun. * * * This little book should be in the hands of all dog-lovers, and we draw the attention of the fair sex particularly to some of the hints given in it.

The Madras Mail.—* * * We have every confidence in recommending his (Major O.'s) book as one which will be found most useful to persons who know how to value, and would wish to keep in health, a good dog.

Calcutta, February 185

THACKER, SPINK AND CO.'S,
PUBLICATIONS.

—CONTENTS—

POETRY, FICTION, ETC.

THE SPOILT CHILD.—A TALE OF HINDU DOMESTIC LIFE. A Translation by G. D. OSWELL, M.A., of the Bengali Novel "*Alaler Gharer Dulal,*" by PRARY CHAND MITTER (Tek Chand Thakur). Crown 8vo, cloth, Rs. 3; paper, Rs. 2-8.

"Interesting as throwing a fairly vivid light upon the intimate life of a Hindu household."—*Daily Chronicle.*

"May be heartily commended both for its literary qualities and for the vivid picture it gives of Bengali manners and customs."—*Scotsman.*

"Every chapter of the story is a picture of native thought and native prejudice, presenting the ever-enduring hatreds of Hindoo and Mussalman in a vivid light."—*Times Weekly.*

"Its merit lies in the quaint humour and quaint illustrations with which the author embellishes his narrative."—*Academy.*

"Mr. Oswell's pleasant translation."—*Athenæum.*

SONG OF SHORUNJUNG & OTHER POEMS.—Crown 8vo, cloth. Rs. 2-8.
CONTENTS:—Darjeeling: Summer—The Song of Shorunjung—The Tsari Reed—To the Uplands—A Pastoral—The Jessamine—The Fakir—The Fisher's Supper—A Son—Two Moods—Farewell to Devon—Song—The London Maid —Infancy—A Lullaby—There are Words—Borodino—The Lone Night—The Captive—Cossack Cradle Song—Gifts of the Terak—The Cup of Life—Scenes from Eugene Onyegin.

"Full of tastefully conceived description. . . . Some of the single verses are very tuneful......A number of translations from the Russian form a noteworthy part of it. The poems have been admirably done into English, the translator having not only retained the sense of the original but the distinctive Russian character of expression and metre."—*Englishman.*

THACKER, SPINK AND CO., CALCUTTA.

RHYMING LEGENDS OF IND.—By H. KIRKWOOD GRACEY, B.A., C.S. Crown 8vo, cloth, Rs. 5-8.

CONTENTS:—The City of Gore—A Mother's Vengeance—The Blue Cow—Famine—A Terrible Tiger—The Legend of Somnath—Treasure Trove—The Idol of Kalinga—Mind *vs.* Matter—*Vultur in partibus.*

"A collection of bright little poems. Keen satirical touches are introduced here and there throughout the volume....The clever little book."—*Morning Post.*

"A charming little book. Of the poems here collected the majority will bear reading several times over. The author writes in lively mirth-provoking fashion."—*Express.*

"The whole volume is, indeed, well worth reading; it is an enjoyable little publication."—*Madras Mail.*

"The writer of the present volume handles his theme with remarkable ability."—*Bookseller.*

ELSIE ELLERTON.—A NOVELETTE OF ANGLO-INDIAN LIFE. BY MAY EDWOOD, author of "Stray Straws," "Autobiography of a Spin," etc. Crown 8vo. Re. 1-8.

'This novel is amusing, pure in tone, and distinguished by much local colouring."—*Athenæum.*

"Rudyard Kipling has taken the most becoming phase of Anglo-Indian life and in the main made the worst of it. Miss Edwood takes the pleasantest and makes the most of it."—*Home News.*

INDIAN LYRICS.—By W. TREGO WEBB, M.A., Professor of English Literature, Presidency College. Fcap. 8vo, cloth. Rs. 4.

"Vivacious and clever . . . . He presents the various sorts and conditions of humanity that comprise the round of life in Bengal in a series of vivid vignettes .. He writes with scholarly directness and finish."—*Saturday Review.*

"A volume of poems of more than ordinary interest and undoubted ability." —*Oxford and Cambridge Undergraduate's Journal.*

LIGHT AND SHADE.—BY HERBERT SHERRING. A Collection of Tales and Poems. Crown 8vo, cloth. Rs. 3.

"Piquant and humorous—decidedly original—not unworthy of Sterne."— *Spectator* (London).

STRAY STRAWS.—BEING A COLLECTION OF SKETCHES AND STORIES. By MIGNON. Crown 8vo. Re. 1-8.

"It is a capital book to take up when one has a few spare moments on hand."—*Englishman.*

"A very interesting collection of short stories and sketches."—*Morning Post* (Allahabad).

THACKER, SPINK AND CO., CALCUTTA.

BARRACK ROOM BALLADS AND OTHER VERSES.—BY RUDYARD KIPLING. Printed by Constable on laid paper, rough edges, bound in buckram, gilt top. Post 8vo, Rs. 6.

"Mr. Kipling's verse is strong, vivid, full of character......unmistakable genius rings in every line."—*Times*.

" 'The finest things of the kind since Macaulay's 'Lays.' "—*Daily Chronicle*.
"Mr. Kipling is probably our best ballad writer since Scott."—*Daily News*.
"One of the books of the year."—*National Observer*.

A QUEER ASSORTMENT.—A COLLECTION OF SKETCHES. BY A. HARVEY JAMES. Crown 8vo. Re. 1.

POPPIED SLEEP.—A CHRISTMAS STORY OF AN UP-COUNTRY STATION By Mrs. H. A. FLETCHER, author of "Here's Rue for You." Crown 8vo, sewed. Re. 1-8.

PLAIN TALES FROM THE HILLS.—BY RUDYARD KIPLING, author of "Departmental Ditties & other Verses." Third Edition. Crown 8vo, Rs. 5.

"Rattling stories of flirtation and sport . . . . Funny stories of practical jokes and sells . . . . Sad little stories of deeper things told with an affect-ation of solemnity but rather more throat-lumping for that."—*Sunday Times*.

"Mr. Kipling possesses the art of telling a story. 'Plain Tales from the Hills' sparkle with fun; they are full of life, merriment, and humour, as a rule mirth-provoking. There is at times a pathetic strain; but this soon passes, and laughter—as the Yankees say, side-splitting laughter—is the order of the day."—*Allen's Indian Mail*.

"It would be hard to find better reading."—*Saturday Review*.

A ROMANCE OF THAKOTE AND OTHER TALES.—Reprinted from *The World, Civil and Military Gazette*, and other Papers. By F. C. C. Crown 8vo. Re. 1.

INDIAN MELODIES. — BY GLYN BARLOW, M.A., Professor, St. George's College, Mussoorie. Fcap. 8vo, cloth. Rs. 2.

"Interesting, pleasant and readable . . . Mr. Barlow's little volume deserves a kindly and favourable reception, and well repays perusal."—*The Morning Post*.

LEVIORA.—BEING THE RHYMES OF A SUCCESSFUL COMPETITOR. By the late T. F. BIGNOLD, Bengal Civil Service. 8vo, sewed. Rs. 2.

THACKER, SPINK AND CO., CALCUTTA.

INDIAN IDYLLS.—BY AN IDLE EXILE. Author of "In Tent and Bungalow" and "by a Himalayan Lake." Crown 8vo, cloth.   Rs. 2-8.

CONTENTS:—The Maharajah's Guest—The Major's Mess Clothes—In a Haunted Grove—How we got rid of Honks—My Wedding Day—Mrs. Caramel's Bow-Wow—The Tables Turned—A Polo Smash—After the Wily Boar —In the Rajah's Palace—Two Strings—A Modern Lochinvar—My First Snipe —Mrs. Dimple's Victim—Lizzie; a Shipwreck—How the Convalescent Depôt killed a Tiger—Faithful unto Death—The Haunted Bungalow—Christmas with the Crimson Cuirassiers—In Death they were not Divided.

"A pleasant little book of short stories and sketches, bright and light for the most part, dealing with frays and feasting, polo and pigsticking, with a ghost story thrown in by way of relief."—*Saturday Review.*

"All these are thoroughly Indian in colour and tone, but are not the less amusing and sprightly matter for reading in idle half hours."—*Daily Telegraph.*

"A series of crisp little stories . . . we shall be surprised if it fails to fetch' the public who have had nothing better to amuse them since the lamented Aberigh Mackay astonished Anglo-India with his Sir Ali Baba's revelations."—*Express.*

THE TRIBES ON MY FRONTIER.—AN INDIAN NATURALIST'S FOREIGN POLICY. By EHA. With 50 Illustrations by F. C. MACRAE. Imp. 16mo. Uniform with "Lays of Ind." Fourth Edition. Rs. 7.

"We have only to thank our Anglo-Indian naturalist for the delightful book which he has sent home to his countrymen in Britain. May he live to give us another such."—*Chambers' Journal.*

"A most charming series of sprightly and entertaining essays on what may be termed the fauna of the Indian Bungalow. . . . . . We have no doubt that this amusing book will find its way into every Anglo-Indian's library."—*Allen's Indian Mail.*

"This is a delightful book, irresistibly funny in description and illustration, but full of genuine science too. . . . . . . There is not a dull or uninstructive page in the whole book."—*Knowledge.*

INDIA IN 1983.—A REPRINT OF THIS CELEBRATED PROPHESY OF NATIVE RULE IN INDIA. Fcap. 8vo. Re. 1.

"Instructive as well as amusing."—*Indian Daily News.*

"There is not a dull page in the hundred and thirty-seven pages of which it consists."—*Times of India.*

REGIMENTAL RHYMES AND OTHER VERSES.—By KENTISH RAG. Imp. 16mo. Sewed, Re. 1; cloth, Rs. 1-8.

"'Kentish Rag' has been well advised to collect his writings"—*Madras Mail.*

"The *verve* and go of the Regimental Rhymes are undoubted."—*Nilgiri News.*

"The rhymes have a thoroughly pleasing heartiness and frank jollity of their own which should certainly recommend them to the gallant Regiments whose deeds they chronicle."—*The Soldier.*

THACKER, SPINK AND CO., CALCUTTA.

AUTOBIOGRAPHY OF A SPIN.—By MAY EDWOOD, author of "Elsie Ellerton," "Stray Straws," &c. Rs. 1-8.

"Undeniably a clever and not unaffecting study of the natural history of flirtation."—*Saturday Review.*

"In Anglo-Indian society slang a 'Spin' appears to be a young lady who visits India in order to obtain matrimony by means of a vivacious, not to say bold, demeanour. The author of this book describes such a person, and her heartless conduct in the pursuit of her object, which, however, does not appear to have been successful."—*Times Weekly.*

LALU, THE CHILD-WIDOW.—A Poem in seven parts : Proem—The Zemindar—The Farm—The Betrothal—The Lovers—Widowhood—The Pyre—Rest. By Lt.-Col. W. L. Greenstreet. Cr. 8vo. Rs. 2.

BEHIND THE BUNGALOW.—By EHA, author of "THE TRIBES ON MY FRONTIER." With Illustrations by F. C. MACRAE. Fourth Edition. Imp. 16mo. Rs. 5.

"Of this book it may conscientiously be said that it does not contain a dull page, while it contains very many which sparkle with a bright and fascinating humour, refined by the unmistakable evidences of culture."—*Home News.*

'The author of 'Behind the Bungalow' has an excellent sense of humour combined with a kindliness of heart which makes his little book delightful reading."—*Saturday Review.*

"There is plenty of fun in 'Behind the Bungalow.'"—*World.*

"A series of sketches of Indian servants, the humour and acute observation of which will appeal to every Anglo-Indian."—*Englishman.*

"Drawn with delightful humour and keen observation."—*Athenæum.*

"Every variety of native character, the individual as well as the nation, caste, trade, or class, is cleverly portrayed in these diverting sketches."—*Illustrated London News.*

INDIAN ENGLISH AND INDIAN CHARACTER.—By ELLIS UNDERWOOD. Fcap. 8vo. As. 12.

LAYS OF IND.—By ALIPH CHEEM. Comic, Satirical, and Descriptive Poems illustrative of Anglo-Indian Life. Eighth Edition. Enlarged. With 70 Illustrations. Cloth, elegant gilt edges. Rs. 7-8.

"There is no mistaking the humour, and at times, indeed, the fun is both 'fast and furious.' One can readily imagine the merriment created round the camp fire by the recitation of 'The Two Thumpers,' which is irresistibly droll. —*Liverpool Mercury.*

"The verses are characterised by high animal spirits, great cleverness, and most excellent fooling."—*World.*

THE CAPTAIN'S DAUGHTER.—A NOVEL. By A. C. POOSHKIN. Literally translated from the Russian by STUART H. GODFREY, Captain, Bo. S. C. Crown 8vo. Rs. 2.

"Possesses the charm of giving vividly, in about an hour's reading, a conception of Russian life and manners which many persons desire to possess."—*Englishman.*

"The story will interest keenly any English reader."—*Overland Mail.*

THACKER, SPINK AND CO., CALCUTTA.

A NATURALIST ON THE PROWL.—By EHA, Author of "Tribes on
my Frontier," "Behind the Bungalow." Imp. 16mo. Rs. 8.
Profusely illustrated by Photographs of Animals in their habitat and at
work; and Sketches by R. A. STERNDALE.

"The Author is a keen observer of nature, and his descriptions are felicitous
and apt. He is entirely at home amid the Indian fauna, and writes of things
which he knows and loves."—*Glasgow Herald.*

"A charming record of wild life in the jungle."—*Saturday Review.*

"Attractive alike in subject, treatment and appearance."—*Literary World.*

"Very easy and delightful reading."—*The Field.*

"We have not for a long time come across so readable a volume as this."
—*Admiralty and Horse Guards Gazette.*

"Anyone who takes up this book will follow our example and not leave his
chair until he has read it through. It is one of the most interesting books
upon natural history that we have read for a long time."—*Daily Chronicle.*

"HERE'S RUE FOR YOU."—NOVELETTES, ENGLISH AND ANGLO-
INDIAN. By Mrs. H. A. FLETCHER. Crown 8vo, sewed. Rs. 2.

CONTENTS:—A Summer Madness—Whom the Gods Love—Nemesis—A
Gathered Rose—At Sea: a P. and O. Story—Esther: an Episode.

ONOOCOOL CHUNDER MOOKERJEE.—A MEMOIR OF THE LATE
JUSTICE ONOOCOOL CHUNDER MOOKERJEE. By M. MOOKERJEE. Fourth
Edition. 12mo. Re. 1.

"The reader is earnestly advised to procure the life of this gentleman,
written by his nephew, and read it."—*The Tribes on my Frontier.*

DEPARTMENTAL DITTIES AND OTHER VERSES.—By RUDYARD
KIPLING. Seventh Edition. With additional Poems. Cloth. Rs. 8.

"This tiny volume will not be undeserving of a place on the bookshelf that
holds 'Twenty-one Days in India.' Its contents, indeed, are not unlike the
sort of verse we might have expected from poor 'Ali Baba' if he had been
spared to give it us. Mr. Kipling resembles him in lightness of touch, quaint-
ness of fancy, and unexpected humour."—*Pioneer.*

"The verses are all written in a light style, which is very attractive, and
no one with the slightest appreciation of humour will fail to indulge in many
a hearty laugh before turning over the last page."—*Times of India.*

"Mr. Kipling's rhymes are rhymes to some purpose. He calls them Departmental Ditties: but they are in reality social sketches of Indian officialism
from a single view point, that of the satirist, though the satire is of the
mildest and most delightful sort."—*Indian Planters' Gazette.*

THE INSPECTOR.—A COMEDY. By GOGOL. Translated from the
Russian by T. HART-DAVIES, Bombay Civil Service. Crown 8vo. Rs. 2.

"His translation, we may add, is a very good one."—*The Academy.*

A MIDSUMMER NIGHT'S DREAM (SHAKESPEARE).—Adapted to Pastoral Representation. By N. NEWNHAM-DAVIS. Crown 8vo. Re. 1.

THACKER, SPINK AND CO., CALCUTTA.

THE SECOND BOMBARDMENT AND CAPTURE OF FORT WILLIAM, CALCUTTA.—An Account of the Bombardment of Fort William, and the Capture and Occupation of the City of Calcutta, on the 20th June 1891, &c., by a Russian Fleet and Army. Compiled from the Diaries of PRINCE SERGH WOHONZOFF and GENERAL YAGODKIN. Translated from the Original Russe by IVAN BATIUSHKA. Crown 8vo, sewed. Re. 1-8.

HISTORY, CUSTOMS, TRAVELS, ETC.

THE ORIGIN OF THE MOHAMEDANS IN BENGAL.—By MOULVIE FUZL RUBBEE. Crown 8vo.

THE HINDOOS AS THEY ARE.—A DESCRIPTION OF THE MANNERS, Customs, and Inner Life of Hindoo Society, Bengal. By SHIB CHUNDER BOSE. Second Edition, Revised. Crown 8vo, cloth. Rs. 5.

HINDU MYTHOLOGY.—VEDIC AND PURANIC. By W. J. WILKINS, of the London Missionary Society, Calcutta. Profusely Illustrated. Imp. 16mo, cloth, gilt elegant. Rs. 7-8.

" His aim has been to give a faithful account of the Hindu deities such as an intelligent native would himself give, and he has endeavoured, in order to achieve his purpose, to keep his mind free from prejudice or theological bias. The author has attempted a work of no little ambition and has succeeded in his attempt, the volume being one of great interest and usefulness."—*Home News.*

" Mr. Wilkins has done his work well, with an honest desire to state facts apart from all theological prepossession, and his volume is likely to be a useful book of reference."—*Guardian.*

MODERN HINDUISM.—BEING AN ACCOUNT OF THE RELIGION AND LIFE of the Hindus in Northern India. By W. J. WILKINS, author of " Hindu Mythology, Vedic and Puranic." Demy 8vo. Rs. 8.

" He writes in a liberal and comprehensive spirit."—*Saturday Review.*

"......volume which is at once a voluminous disquisition upon the Hindu religion, and a most interesting narrative of Hindu life, the habits and customs of the Hindu community and a national Hindu historiette, written with all the nerve of the accomplished littérateur, added to the picturesque word-painting and life-like delineations of a veteran novelist."—*Lucknow Express.*

" A solid addition to our literature."—*Westminster Review.*

" A valuable contribution to knowledge."—*Scotsman.*

THE DHAMMAPADA; OR, SCRIPTURAL TEXTS. A Book of Buddhist Precepts and Maxims. Translated from the Pali on the Basis of Burmese Manuscripts. By JAMES GRAY. Second Edition. 8vo, boards. Rs. 2.

THE ETHICS OF ISLAM.—A LECTURE BY THE HON'BLE AMEER ALI, C.I.E., author of " The Spirit of Islam," " The Personal Law of the Mahommedans," etc. Crown 8vo. Cloth gilt. Rs. 2-8.

An attempt towards the exposition of Islâmic Ethics in the English language. Besides most of the Koranic Ordinances, a number of the precepts and sayings of the Prophet, the Caliph Ali, and of 'Our Lady,' are translated and given.

THACKER, SPINK AND CO., CALCUTTA.

THE LIFE AND TEACHING OF KESHUB CHUNDER SEN.—BY P. C. MASUMDAR. Second and Cheaper Edition. Rs. 2.

THEOSOPHICAL CHRISTIANITY.—AN ADDRESS BY L. S. Second Edition. Revised and Enlarged. Small 4to. As. 8.

KASHGARIA (EASTERN OR CHINESE TURKESTAN).—HISTORICAL, Geographical, Military, and Industrial. By COL. KUROPATKIN, Russian Army. Translated by Maj. GOWAN, H. M.'s Indian Army. 8vo. Rs. 2.

ANCIENT INDIA AS DESCRIBED BY PTOLEMY.—WITH INTRODUCTION, Commentary, Map of India. By J. W. MCCRINDLE, M.A. 8vo, cloth, lettered. Rs. 4-4.

ANCIENT INDIA AS DESCRIBED BY MEGASTHENES AND ARRIAN. With Introduction, Notes, and a Map of Ancient India. By J. W. MCCRINDLE, M.A. 8vo. Rs. 2-8.

THE COMMERCE AND NAVIGATION OF THE ERYTHRÆAN SEA: Periplus Maris Erythræi; and of Arrian's Account of the Voyage of Nearkhos. With Introduction, Commentary, Notes, and Index. By J. W. MCCRINDLE, M.A. 8vo. Rs. 3.

ANCIENT INDIA AS DESCRIBED BY KTESIAS THE KNIDIAN.— A Translation of the Abridgment of his 'Indika,' by Photios. With Introduction, Notes, Index. By J. W. MCCRINDLE. M.A. 8vo. Rs. 3.

A MEMOIR OF CENTRAL INDIA, INCLUDING MALWA AND ADJOINING PROVINCES, with the History, and copious Illustrations, of the Past and Present condition of that country. By Maj.-Gen. S. J. MALCOLM, G.C.B., &c. *Reprinted from Third Edition.* 2 vols. Crown 8vo, cloth. Rs. 5.

BOOK OF INDIAN ERAS.—WITH TABLES FOR CALCULATING INDIAN DATES. By ALEXANDER CUNNINGHAM, C.S.I., C.I.E., Major-General, Royal Engineers. Royal 8vo, cloth. Rs. 12.

TALES FROM INDIAN HISTORY.—BEING THE ANNALS OF INDIA retold in Narratives. By J. TALBOYS WHEELER. Crown 8vo, cloth, Rs. 3. School Edition, cloth, limp, Re. 1-8.

"The history of our great dependency made extremely attractive reading. Altogether this is a work of rare merit."—*Broad Arrow.*

"Will absorb the attention of all who delight in thrilling records of adventure and daring. It is no mere compilation, but an earnest and brightly written book."—*Daily Chronicle.*

A CRITICAL EXPOSITION OF THE POPULAR "JIHAD."—Showing that all the Wars of Mahammad were defensive, and that Aggressive War or Compulsory Conversion is not allowed in the Koran, &c. By Moulavi CHERAGH ALI. author of "Reforms under Moslem Rule," "Hyderabad under Sir Salar Jung." 8vo. Rs. 6.

MAYAM-MA: THE HOME OF THE BURMAN.—BY TSAYA (REV. H. POWELL). Crown 8vo. Rs. 2.

AN INTRODUCTION TO THE STUDY OF HINDUISM.—BY GURU PRASHAD SEN. Crown 8vo, cloth Rs. 3; paper Rs. 2.

THACKER, SPINK AND CO., CALCUTTA.

CHIN-LUSHAI LAND. INCLUDING A DESCRIPTION OF THE VARIOUS Expeditions into the Chin Lushai Hills and the Final Annexation of the Country. By Surgn.-Lieut.-Colonel A. S. REID, M.B., Indian Medical Service. With three Maps and eight Phototint Illustrations. Cloth, gilt. Rs. 12.

Contains a description of the Chin-Lushai Hills and their inhabitants, from the earliest records, with an account of the various expeditions into the country, the last, *viz.*, that of 1889-90, which led to the final annexation of the wild mountainous tract which lies between India and Burma, being given in full detail.

"A valuable contribution to the History of our Indian border."—*Pall Mall Gazette.*

THE IMAGE OF WAR, OR SERVICE IN THE CHIN HILLS.—A COLLECTION of 34 Full-page Collotypes of Instantaneous Photographs and 160 interspersed in the reading. By Surgeon-Captain A. G. NEWLAND. With Introductory Notes, by J. D. MACNABB, Esq., B.C.S. 4to, cloth, gilt elegant. Rs. 32.

"It would be difficult to give a more graphic picture of an Indian Frontier Expedition."—*The Times.*

"Brings home to us the Chins themselves, their ways and homes, the nature of the country marched through, method of campaign, and daily social habits and experiences of the campaigners. The letterpress, bright and simple, is worthy of the photographs."—*Broad Arrow.*

"The pictures are so good and spirited and so well reproduced that we are apt to forget that the letterpress is also well worth studying."—*Daily Telegraph.*

THE RACES OF AFGHANISTAN.—BEING A BRIEF ACCOUNT of the principal Nations inhabiting that country. By Surg.-Maj. H. W. BELLEW, C.S.I., late on Special Political Duty at Kabul. 8vo, cloth. Rs. 2.

KURRACHEE ; ITS PAST, PRESENT, AND FUTURE.—BY ALEXANDER F. BAILLIE, F.R.G.S., author of "A Paraguayan Treasure," &c. With Maps, Plans, and Photographs, showing the most recent improvements. Super-royal 8vo, cloth. Rs. 15.

THE TRIAL OF MAHARAJA NANDA KUMAR.—A NARRATIVE OF A JUDICIAL MURDER. By H. BEVERIDGE, B.C.S. Demy 8vo. Rs. 5.

"Mr. Beveridge has given a great amount of thought, labour, and research to the marshalling of his facts, and he has done his utmost to put the exceedingly complicated and contradicting evidence in a clear and intelligible form."—*Home News.*

THE EMPEROR AKBAR.—A CONTRIBUTION TOWARDS THE HISTORY OF INDIA in the 16th Century. By FREDERICK AUGUSTUS, Count of Noer, Translated from the German by ANNETTE S. BEVERIDGE. 2 vols. 8vo. cloth, gilt. Rs. 5.

THACKER, SPINK AND CO., CALCUTTA.

ECHOES FROM OLD CALCUTTA.—BEING CHIEFLY REMINISCENCES of the days of Warren Hastings, Francis, and Impey. By H. E. BUSTEED. Second Edition, Enlarged and Illustrated. Post 8vo. Rs. 6.

"The book will be read by all interested in India."—*Army & Navy Magazine.*

"Dr. Busteed's valuable and entertaining 'Echoes from Old Calcutta' has arrived at a second edition, revised, enlarged and illustrated with portraits and other plates rare or quaint. It is a pleasure to reiterate the warm commendation of this instructive and lively volume which its appearance called forth some years since."—*Saturday Review.*

"A series of illustrations which are highly entertaining and instructive of the life and manners of Anglo-Indian society a hundred years ago . . . The book from first to last has not a dull page in it, and it is a work of the kind of which the value will increase with years."—*Englishman.*

CAPT. HAYES' WORKS ON HORSES.

ON HORSE BREAKING.—BY CAPTAIN M. H. HAYES. Numerous Illustrations by J. H. OSWALD-BROWN. Square. Rs. 16.

(1) Theory of Horse Breaking. (2) Principles of Mounting. (8) Horse Control. (4) Rendering Docile. (5) Giving Good Mouths. (6) Teaching to Jump. (7) Mount for the First Time. (8) Breaking for Ladies' Riding. (9) Breaking to Harness. (10) Faults of Mouth. (11) Nervousness and Impatience. (12) Jibbing. (13) Jumping Faults. (14) Faults in Harness. (15) Aggressiveness. (16) Riding and Driving Newly-Broken Horse. (17) Stable Vices.

"One great merit of the book is its simplicity."—*Indian Daily News.*

"A work which is entitled to high praise at being far and away the best reasoned-out one on breaking under a new system we have seen."—*Field.*

"Clearly written."—*Saturday Review.*

"The best and most instructive book of its class that has appeared for many years."—*Times of India.*

RIDING: ON THE FLAT AND ACROSS COUNTRY.—A GUIDE TO PRACTICAL HORSEMANSHIP. By Captain M. H. HAYES. With 70 Illustrations by STURGESS and J. H. OSWALD-BROWN. Third Edition, Revised and Enlarged. Rs. 7-8.

The whole text has been so revised or re-written as to make the work the most perfect in existence, essential to all who wish to attain the art of riding correctly.

"One of the most valuable additions to modern literature on the subject."—*Civil and Military Gazette.*

"A very instructive and readable book."—*Sport.*

"This useful and eminently practical book."—*Freeman's Journal.*

THE POINTS OF THE HORSE.—A FAMILIAR TREATISE ON EQUINE CONFORMATION. Describing the points in which the perfection of each class of horses consists. By Captain M. H. HAYES. Illustrated.

[*New Edition in the Press.*

THACKER, SPINK AND CO., CALCUTTA.

INDIAN RACING REMINISCENCES.—BEING ENTERTAINING NARRA-
TIVES, and Anecdotes of Men, Horses, and Sport. By Captain M. H.
HAYES. Illustrated with 42 Portraits and Engravings. Imp. 16mo. Rs. 6.
" Captain Hayes has done wisely in publishing these lively sketches of life
in India. The book is full of racy anecdote."—*Bell's Life.*

" All sportsmen who can appreciate a book on racing, written in a chatty
style and full of anecdote, will like Captain Hayes' latest work."—*Field.*

" Many a racing anecdote and many a curious character our readers will
find in the book, which is very well got up, and embellished with many
portraits."—*Bailey's Magazine.*

VETERINARY NOTES FOR HORSE-OWNERS.—A POPULAR GUIDE
to Horse Medicine and Surgery. By Captain M. H. HAYES. Fourth Edition, Enlarged and Revised to the latest Science of the Day. With many
New Illustrations by J. H. OSWALD-BROWN. Crown 8vo, buckram.
[*New Edition in the Press.*

The chief new matter in this Edition is—Articles on Contracted Heels,
Donkey's Foot Disease, Forging or Clicking, Rheumatic Joint Disease,
Abscess, Dislocation of the Shoulder Joint, Inflammation of the Mouth and
Tongue, Flatulent Distention of the Stomach, Twist of the Intestines, Relapsing
Fever, Cape Horse Sickness, Horse Syphilis, Rabies, Megrims, Staggers,
Epilepsy, Sunstroke, Poisoning, Castration by the Ecraseur, and Mechanism
of the Foot (in Chapter or Shoeing).

" Of the many popular veterinary books which have come under our notice,
this is certainly one of the most scientific and reliable. . . . . The description of symptoms and the directions for the application of remedies are
given in perfectly plain terms, which the tyro will find no difficulty in com-
prehending."—*Field.*

" Simplicity is one of the most commendable features in the book."—*Illustrated Sporting and Dramatic News.*

"Captain Hayes, in the new edition of 'Veterinary Notes,' has added considerably to its value, and rendered the book more useful to those non-professional people who may be inclined or compelled to treat their own horses
when sick or injured."—*Veterinary Journal.*

" We do not think that horse-owners in general are likely to find a more
reliable and useful book for guidance in an emergency."—*Field.*

TRAINING AND HORSE MANAGEMENT IN INDIA.—BY CAPTAIN M.
H. HAYES, author of " Veterinary Notes for Horse-Owners," " Riding," &c.
Fifth Edition. Crown 8vo. Rs. 6.

" No better guide could be placed in the hands of either amateur horseman
or veterinary surgeon."—*Veterinary Journal.*

" A useful guide in regard to horses anywhere. . . . Concise, practical,
and portable."—*Saturday Review.*

THACKER, SPINK AND CO., CALCUTTA.

SOUNDNESS AND AGE OF HORSES.—A VETERINARY AND LEGAL
GUIDE to the Examination of Horses for Soundness. By Captain M. H.
HAYES, M.R.C.V.S. With 100 Illustrations. Crown 8vo. Rs. 6.

"Captain Hayes is entitled to much credit for the explicit and sensible
manner in which he has discussed the many questions—some of them extremely vexed ones—which pertain to soundness and unsoundness in horses."
—*Veterinary Journal.*

"All who have horses to buy, sell, or keep will find plenty to interest them
in this manual, which is full of illustrations, and still fuller of hints and
wrinkles."—*Referee.*

"Captain Hayes' work is evidently the result of much careful research, and
the horseman, as well as the veterinarian, will find in it much that is interesting and instructive."—*Field.*

THE HORSE-WOMAN.—A PRACTICAL GUIDE TO SIDE-SADDLE RIDING.
By Mrs. HAYES, and Edited by Captain M. H. HAYES. Illustrated by 48
Drawings by J. OSWALD-BROWN and 4 Photographs. Uniform with
"Riding: on the Flat and Across Country." Imp. 16mo. Rs. 7-8.

"This is the first occasion on which a practical horseman and a practical
horsewoman have collaborated in bringing out a book on riding for ladies.
The result is in every way satisfactory."—*Field.*

"A large amount of sound practical instruction, very judiciously and plea-
santly imparted."—*Times.*

"We have seldom come across a brighter book than 'The Horsewoman.'"—
Athenæum.

"Eminently sensible and practical."—*Daily Chronicle.*

SPORT AND VETERINARY WORKS.

HIGHLANDS OF CENTRAL INDIA.—NOTES ON THEIR FORESTS
and Wild Tribes, Natural History, and Sports. By Capt. J. FORSYTH, B.S.C.
New Edition. With Map and Tinted Illustrations. Rs. 7-8.

CALCUTTA TURF CLUB RULES OF RACING, together with the
Rules relating to Lotteries, Betting, Defaulters, and the Rules of the
Calcutta Turf Club. Revised May 1892. Authorized Edition. Rs. 2.

THE RACING CALENDAR, VOL. VI, FROM MAY 1893 TO APRIL 1894,
RACES PAST. Published by the Calcutta Turf Club. CONTENTS:—Rules
of Racing, Lotteries, C. T. C., etc., Registered Colours; Licensed Train-
ers and Jockeys; Assumed Names; List of Horses Aged, Classed and
Measured by C. T. C. and W. I. T. C.; Races Run under C. T. C. Rules;
Performances of Horses; Appendix and Index. Rs. 4.

THE RACING CALENDAR FROM 1ST AUGUST 1888 TO 30TH APRIL 1889,
RACES PAST. 12mo, cloth. Vol. I, Rs. 4. Vol. II, to April 1890,
Rs. 4. Vol. III, to April 1891, Rs. 4. Vol. IV, to April 1892, Rs. 4.
Vol. V, to April 1893, Rs. 4.

THACKER, SPINK AND CO., CALCUTTA.

CALCUTTA RACING CALENDAR.—PUBLISHED EVERY FORTNIGHT.
Annual Subscription, Rs. 12.

THE SPORTSMAN'S MANUAL.—IN QUEST OF GAME IN KULLU,
Lahoul, and Ladak to the Tso Morari Lake, with Notes on Shooting in
Spiti, Bara Bagahal, Chamba, and Kashmir, and a detailed description of
Sport in more than 100 Nalas. With 9 Maps. By Lt.-Col. R. H. TYACKE,
late H. M.'s 98th and 34th Regiments. Fcap. 8vo, cloth. Rs. 8-8.

SEONEE : OR, CAMP LIFE ON THE SATPURA RANGE.—A Tale of
Indian Adventure. By R. A. STERNDALE, author of "Mammalia of
India," "Denizens of the Jungles." Illustrated by the author. With a
Map and an Appendix containing a brief Topographical and Historical
Account of the District of Seonee in the Central Provinces of India.
Crown 8vo, cloth. Rs. 7.

LARGE GAME SHOOTING IN THIBET, THE HIMALAYAS, NORTHERN
AND CENTRAL INDIA. By Brig.-General ALEXANDER A. KINLOCH. Containing Descriptions of the Country and of the various Animals to be
found ; together with extracts from a journal of several years' standing.
With 86 Illustrations from photographs and a Map. Third Edition, Revised and Enlarged. Demy 4to, elegantly bound. Rs. 25.

"This splendidly illustrated record of sport, the photogravures, especially
the heads of the various antelopes, are life-like ; and the letter-press is very
pleasant reading."—*Graphic.*

"The book is capitally got up, the type is better than in former editions, and
the excellent photogravures give an exceptional value to the work."—*Asian.*

DENIZENS OF THE JUNGLES.—A SERIES OF SKETCHES OF WILD
ANIMALS, illustrating their form and natural attitude. With Letter-press
Description of each Plate. By E. A. STERNDALE, F.R.G.S., F.Z.S.,
author of "Natural History of the Mammalia of India," "Seonee," &c.
Oblong folio. Rs. 10.

LORD WILLIAM BERESFORD ; SOLDIER, STATESMAN AND SPORTS-
MAN. Being a record of his Eighteen years in India and a full résumé of
his Racing Career. By H. E. ABBOTT. Crown 8vo. Sewed. Re. 1.

REMINISCENCES OF TWENTY YEARS' PIG-STICKING IN BENGAL.
By RAOUL. Illustrated with 6 Portraits. Crown 8vo, cloth gilt.
Rs. 6-12.

HORSE BREEDING AND REARING IN INDIA. —WITH NOTES on
TRAINING for the Flat and Across Country, and on Purchase, Breaking in
and General Management. By Major JOHN HUMFREY, B.S.C., F.E.S.
Crown 8vo, cloth. Rs. 8-8.

THACKER, SPINK AND CO., CALCUTTA.

INDIAN HORSE NOTES.—An Epitome of useful Information arranged for ready reference on Emergencies, and specially adapted for Officers and Mofussil Residents. All Technical Terms explained and Simplest Remedies selected. By Major C——, author of "Indian Notes about Dogs." Third Edition, Revised and considerably Enlarged. Fcap. 8vo, cloth. Rs. 2.

DOGS FOR HOT CLIMATES.—A Guide for Residents in Tropical Countries as to suitable Breeds, their Respective Uses, Management and Doctoring. By Vero Shaw and Captain M. H. Hayes. With Illustrations. [*In the Press.*

HOW TO CHOOSE A DOG, and how to select a Puppy, together with a few notes upon the peculiarities and characteristics of each Breed. By Vero Shaw, author of "The Illustrated Book of the Dog," late Kennel Editor of the "*Field.*"
This small book contains the result of many years' careful study of all breeds of Dogs, and condenses into a small compass a very large amount of information of a kind that has never yet been presented to the public. Forty-five different Breeds of Dogs are described; the compressed details as to description, use, weights at various ages, etc., being given in two pages for each Breed.

GUIDE TO EXAMINATION OF HORSES, for Soundness for Students and Beginners. By Moore, f.r.c.v.s., Army Vety. Dept., Vety. Officer, Remount Depôt, Calcutta. Fcap. 8vo. Limp cloth. Re. 1.

RIDING FOR LADIES, WITH HINTS ON THE STABLE.—A Lady's House Book. By Mrs. Power O'Donoghue. With 75 Illustrations by A. Chantrey Corbould. Elegantly printed and bound. Imp. 16mo, gilt. Rs. 7-8.

INDIAN NOTES ABOUT DOGS.—Their Diseases and Treatment. By Major C——. Fourth Edition. Fcap. 8vo, cloth. Re. 1-8.

ANGLING ON THE KUMAUN LAKES.—With a Map of the Kumaon Lake Country and Plan of each Lake. By Depy. Surgeon-General W. Walker. Crown 8vo, cloth. Rs. 4.
"Written with all the tenderness and attention to detail which characterise the followers of the gentle art."—*Hayes' Sporting News.*

USEFUL HINTS TO YOUNG SHIKARIS ON THE GUN AND RIFLE.—By "The Little Old Bear." Reprinted from *Asian.*

THE ARMS ACT (XI OF 1878).—With all the Notices of the Government of India, the Bengal, North-Western Provinces and Punjab Governments, and High Court Decisions and Rulings. By W. Hawkins. Second Edition. 8vo, cloth. Rs. 7-8.

POLO RULES.—Rules of the Calcutta Polo Club and of the Indian Polo Association, with the Article on Polo by "An Old Hand." Reprinted from *Hayes' Sporting News.* Fcap. 8vo. Re. 1.

THACKER, SPINK AND CO., CALCUTTA.

LAWS OF FOOTBALL (ASSOCIATION).—Containing the Law of the Game—Definition of Terms—Hints to Referees. For Pocket. As. 4.

THE POLO CALENDAR.—COMPILED BY THE INDIAN POLO ASSOCIATION. Contents: Committee of Stewards, Rules for the Regulation of Tournaments, &c.—Rules of the Game—Station Polo—List of Members—List of Existing Polo Ponies, names and description, with Alphabetical List—Records of Tournaments.— Previous Winners, VOL.I, 1892-93. VOL II, 1893-94. Each Re. 1-8.

RULES OF POLO.—From the Polo Calendar. Revised 1894. As. 8.

MEDICINE, HYGIENE, ETC.

AIDS TO PRACTICAL HYGIENE. BY J. C. BATTERSBY, B.A., M.B., B.CH., Univ. Dublin, Fcap. 8vo, cloth. Rs. 2.
"A valuable handbook to the layman interested in sanitation."—*Morning Post.*
"To the busy practitioner or the medical student it will serve the purposes of a correct and intelligent guide."—*Medical Record.*

HINTS FOR THE MANAGEMENT AND MEDICAL TREATMENT OF CHILDREN IN INDIA. By EDWARD A. BIRCH, M.D., Late Principal, Medical College, Calcutta. Third Edition, Revised. Being the Ninth Edition of "Goodeve's Hints for the Management of Children in India." Crown 8vo, cloth. Rs. 7.
The Medical Times and Gazette, in an article upon this work and Moore's "Family Medicine for India," says:—"The two works before us are in themselves probably about the best examples of medical works written for non-professional readers. The style of each is simple, and as free as possible from technical expressions. The modes of treatment recommended are generally those most likely to yield good results in the hands of laymen; and throughout each volume the important fact is kept constantly before the mind of the reader, that the volume he is using is but a poor substitute for personal professional advice, for which it must be discarded whenever there is the opportunity."

THE INDIGENOUS DRUGS OF INDIA. SHORT DESCRIPTIVE NOTICES of the principal Medicinal Products met with in British India. By RAI BAHADUR KANNY LALL DEY, C.I.E., Second Edition. Revised and entirely Re-written. Demy 8vo. [*In the Press.*

QUERIES AT A MESS TABLE.—WHAT SHALL WE EAT? WHAT SHALL WE DRINK? By Surg.-Maj. JOSHUA DUKE. Fcap. 8vo, cloth, gilt. Rs. 2-4.

BANTING IN INDIA.—WITH SOME REMARKS ON DIET AND THINGS in General. By Surg.-Maj. JOSHUA DUKE. Third Edition. Cloth. Re. 1-8.

OUTLINES OF MEDICAL JURISPRUDENCE FOR INDIA.—BY J. D. B. GRIBBLE. M.C.K. (Retired), and PATRICK HEHIR, M.D., F.R.C.S.E. Third Edition, Revised, Enlarged, and Annotated. Demy 8vo. Rs. 5-8.

THACKER, SPINK AND CO., CALCUTTA.

RUDIMENTS OF SANITATION.—FOR INDIAN SCHOOLS. BY PATRICK HEHIR, M.D. Second Edition. 12mo, cloth. Re. 1-4.

THE TEETH.—THEIR STRUCTURE, DISEASE, AND PRESERVATION. With some Notes on Conservative and Prosthetic Dentistry. Nine Plates. By JOS. MILLER, L.D.S., R.C.S.E. Second Edition. 8vo, cloth. Rs. 2-8.

THE BABY.—NOTES ON THE FEEDING, REARING AND DISEASES OF INFANTS. By S. O. MOSES, Licentiate of the Royal College of Physicians, Edinburgh, &c. Fcap. 8vo, cloth. Rs. 2.

MY LEPER FRIENDS.—AN ACCOUNT OF PERSONAL WORK AMONG LEPERS, and their daily life in India. By Mrs. HAYES. With Illustrations from Photographs, and a Chapter on Leprosy by Dr. G. G. MACLAREN. Imp. square 32mo. Rs. 2-8.

" The author pictures a very sad phase of human misery by relating the story of the inner life of sufferers whom she has known."—*Cork Constitution.*

" It is impossible to read Mrs. Hayes' book without feeling the keenest sympathy with her in her brave and onerous work, and it cannot fail to result in a considerable return for the advantage of the lepers. Mrs. Hayes writes well and vividly, and there is a note of thorough sincerity in all she says that lends an additional charm to the work. . . . There are several illustrations in the book, reproduced from photographs of lepers."—*Home News.*

" On the whole, Mrs. Hayes has written her book in a very sympathising spirit."—*Indian Daily News.*

HYGIENE OF WATER AND WATER SUPPLIES.—BY PATRICK HEHIR, M.D., Lecturer on Hygiene, Hyderabad Med. School. 8vo, cloth, flush. Rs. 2.

CHOLERA EPIDEMIC IN KASHMIR, 1892.—BY A. MITRA, L.R.C.P., L.R.C.S., Principal Medical Officer in Kashmir. With Map and Tables. 4to, sewed. Re. 1.

A RECORD OF THREE YEARS' WORK OF THE NATIONAL ASSO-CIATION for Supplying Female Medical Aid to the Women of India, August 1885 to August 1888. By H. E. THE COUNTESS OF DUFFERIN. Crown 8vo. Re. 1.

THE NATIONAL ASSOCIATION FOR SUPPLYING FEMALE MEDI-CAL AID to the Women of India. By H. E. THE COUNTESS of DUFFERIN. Reprinted from the *Asiatic Quarterly Review*, by permission. As. 6.

THE INDIAN MEDICAL SERVICE.—A GUIDE FOR INTENDED CANDI-DATES for Commissions and for the Junior Officers of the Service. By WILLIAM WEBB, M.B., Surgeon, Bengal Army, late Agency Surgeon at the Court of Bikanir, Superintendent of Dispensaries, Jails, and Vaccination in the Bikanir State, and for some time Guardian to H. H. the Maharajah. Crown 8vo. Rs. 4.

" We recommend the book to all who think of competing for admission into the Indian Medical Service."—*Lancet.*

THACKER, SPINK AND CO., CALCUTTA.

**THE CARLSBAD TREATMENT FOR TROPICAL AILMENTS, AND HOW
TO CARRY IT OUT IN INDIA.** By Surgn.-Captain L. TANLETON YOUNG.
Ex. fcap. 8vo. Rs. 4.

"A book not only most useful and most instructive, but very readable and
interesting. It is short . . . . . . because it is pithy. The subjects are
thoroughly and fully treated : we feel no lack, nothing unexplained : but it is
done in a clear and concise style. Every word to the point."—*The Pioneer.*

"The book is of a most useful nature, and inspires confidence by the
candour and fulness of its information and points of guidance."—*Irish Times.*

"The book contains the result of six years' practical experience, and should
be of as much advantage to Medical men as to sufferers."—*Home News.*

PERSONAL AND DOMESTIC HYGIENE FOR THE SCHOOL AND HOME,
being a Text-book on Elementary Physiology, Hygiene, Home Nursing
and First Aid to the Injured ; for Senior Schools and Family Reference.
By Mrs. HAROLD HENDLEY. Medallist National Health Society, England. 36 Illustrations. Ex. fcap. 8vo, cloth. Rs. 2 ; or cloth gilt. Rs. 2-8.

"We are decidedly of opinion that it is the most practical and useful book
of its kind which has been published in India. We trust it will gain a large
circulation in the schools and homes of India."—*Indian Medical Gazette.*

"We can recommend this volume without hesitation. In the absence
of the doctor one might obtain hints from any page of it on Hygiene, Nursing,
Accidents and Emergencies. So far as we can see nothing is omitted, and every
direction is given in simple intelligible language."—*Statesman.*

AGUE ; OR, INTERMITTENT FEVER.—By M. D. O'CONNEL, M.D.
8vo, sewed. Rs. 2.

THE LANDMARKS OF SNAKE-POISON LITERATURE.—Being a
Review of the more important Researches into the Nature of Snake-Poisons.
By VINCENT RICHARDS, F.R.C.S. ED., &c., Civil Medical Officer of
Goalundo, Bengal. Rs. 2-8.

MALARIA ; ITS CAUSE AND EFFECTS : MALARIA AND THE SPLEEN ;
Injuries of the Spleen ; An Analysis of 39 cases. By E. G. RUSSELL, M.R.,
B.S.C. 8vo, cloth. Rs. 8.

MEDICAL JURISPRUDENCE FOR INDIA.—By J. B. LYON, F.C.S., F.C.
Brigade-Surgeon, Professor of Medical Jurisprudence, Grant Medical
College, Bombay. The Legal Matter revised by J. D. INVERARITY, Bar.-
at-law. Second Edition. Illustrated. 8vo. Rs. 16.

INDIAN MEDICAL GAZETTE.—Published Monthly. Subscription
Rs. 18 yearly.

DOMESTIC BOOKS.

THE INDIAN COOKERY BOOK.—A PRACTICAL HANDBOOK TO THE
KITCHEN IN INDIA, adapted to the Three Presidencies. Containing Original
and Approved Recipes in every department of Indian Cookery ; Recipes
for Summer Beverages and Home-made Liqueurs ; Medicinal and other
Recipes ; together with a variety of things worth knowing. By a Thirty-
five Years' Resident. Rs. 3.

THACKER, SPINK AND CO., CALCUTTA.

B

INDIAN DOMESTIC ECONOMY AND RECEIPT BOOK. WITH HINDU-STANEE romanized names. Comprising numerous directions for plain wholesome Cookery, both Oriental and English; with much miscellaneous matter, answering all general purposes of reference connected with Household affairs likely to be immediately required by Families, Messes, and private individuals residing at the Pre-idencies or Out-Stations. By Dr. R. Riddell. Eighth Edition. Revised. Post 8vo, cloth. Rs. 6.

FIRMINGER'S MANUAL OF GARDENING FOR INDIA.—A New Edition (the fourth) thoroughly Revised and Re-written. With many Illustrations. By H. St. J. Jackson. Imp. 16mo, cloth, gilt. Rs. 10.

POULTRY-KEEPING IN INDIA. A SIMPLE AND PRACTICAL BOOK on their care and treatment, their various breeds, and the means of rendering them profitable. By Isa Twedd, author of "Cow-Keeping in India." With Illustrations. Crown 8vo, cloth, gilt. Rs. 4.

COW-KEEPING IN INDIA.—A SIMPLE AND PRACTICAL BOOK on their care and treatment, their various breeds, and the means of rendering them profitable. By Isa Twedd. With 37 Illustrations of the various Breeds, &c. Crown 8vo, cloth, gilt. Rs. 4-8.

"A most useful contribution to a very important subject, and we can strongly recommend it."—*Madras Mail.*

"A valuable contribution to Agricultural Literature in the East."—*Ceylon Observer.*

ENGLISH ETIQUETTE FOR INDIAN GENTLEMEN.—By W. Trego Webb, Bengal Educational Department. Second Edition. Fcap. 8vo, cloth. Re. 1-4; paper, Re. 1.

The book comprises chapters on General Conduct, Calls, Dining-out, Levées, Balls, Garden-parties, Railway-travelling, &c. It also contains a chapter on Letter-writing, proper Modes of Address, &c., together with hints on how to draw up Applications for Appointments, with Examples.

PERSONAL AND DOMESTIC HYGIENE FOR THE SCHOOL AND HOME; being a Text-book on Elementary Physiology, Hygiene, Home Nursing, and First Aid to the Injured; for Senior Schools and Family Reference. By Mrs. Harold Hendley. Ex. fcap. 8vo, cloth. Rs. 2; or cloth gilt. Rs. 2-8.

THE AMATEUR GARDENER IN THE HILLS.—With a few Hints on Fowls, Pigeons, and Rabbits. By an Amateur. Second Edition, Revised and Enlarged. Crown 8vo. Rs. 2-8.

FLOWERS AND GARDENS IN INDIA.—A MANUAL FOR BEGINNERS By Mrs. R. Temple-Wright. Third Edition.   [*In the Press.*

......"A most useful little book which we cannot too strongly recommend. We can recommend it to our readers with the utmost confidence, as being not only instructive, but extremely interesting, and written in a delightfully easy, chatty strain."—*Civil and Military Gazette.*

"Very practical throughout. There could not be better advice than this, and the way it is given shows the enthusiasm of Mrs. Temple-Wright." —*Pioneer.*

"It is written in a light conversational style, and it imparts instruction very clearly."—*The Englishman.* ———

THACKER, SPINK AND CO., CALCUTTA.

HINTS FOR THE MANAGEMENT AND MEDICAL TREATMENT OF CHILDREN IN INDIA. By EDWARD A. BIRCH, M.D., late Principal, Medical College, Calcutta. Third Edition, Revised and Enlarged. Being the Ninth Edition of "Goodeve's Hints for the Management of Children in India." Crown 8vo. Rs. 7.

QUERIES AT A MESS TABLE.—WHAT SHALL WE EAT? WHAT SHALL WE DRINK? By Surg.-Maj. JOSHUA DUKE. Fcap. 8vo. cloth, gilt, Rs. 2-4.

THE MEM-SAHIB'S BOOK OF CAKES, BISCUITS, ETC.—With Remarks on Ovens, & Hindustani Vocabulary, Weights & Measures. 18mo, cloth, Rs. 2.

GUIDE BOOKS.

INCE'S KASHMIR HANDBOOK.—A GUIDE FOR VISITORS. Re-Written and much Enlarged by JOSHUA DUKE, Surg.-Lt.-Col., Bengal Medical Service, formerly Civil Surgeon, Gilgit and Srinagar. Fcap. 8vo, cloth, Maps in cloth case. With Appendix containing the Jhelum Valley Road. Rs. 6-8.

The CHIEF CONTENTS are:—An Account of the Province of Kashmir, its Rivers, Lakes, Mountains, Vales, Passes, Inhabitants—Srinagar—Arts and Manufactures, Antiquities, etc.—Requisites for the Journey—Cost—Official Notification to Travellers—Useful Hints—Routes, Gujrat and Pir Panjal—Jhelum, Tangrot and Kotli Poonch—Rawal Pindi and Murree—The New Road—Husan Abbal, Abbottabad, the Jhelum—The Kishengunga Valley—Eastern Portion of Kashmir—Leh—Western Portion of Kashmir—Woolar Lake—Gulmarg—Lolab Valley, Ladak—Pangkong Lake—Gilgit—Astor—Skardu—The Tilaib Valley, &c., and the following—

MAPS:—(1) Jammu and Kashmir with adjoining countries. (2) Map showing Routes to Skardu, etc. (3) Map showing Leh to Himis Monastery, Salt Lake Valley, Pangkong Lake, Kamri Pass, Burail Pass. (4) Astor and Gilgit with surrounding country. *The Maps are finely executed by the Survey of India Dept.*

RAWAL PINDI TO SRINAGAR.—A DETAILED ACCOUNT OF THE NEW Jhelum Valley Road; together with a Brief Note of five other Routes leading into the Valley. Being an Appendix to Ince's Handbook to Kashmir. Re. 1-8.

FROM SIMLA TO SHIPKI IN CHINESE THIBET.—An Itinerary of the Roads and various minor Routes, with a few Hints to Travellers, and Sketch Map. By Major W. F. GORDON-FORBES, Rifle Brigade. Fcap. 8vo, cloth. Rs. 2.

ITINERARIES—Simla to Shipki, 'Charling' Pass, 'Sarahan to Narkunda Forest Road, Simla to the 'Chor,' Pooi to Dankar, Chini to Landour, and the 'Shalle.'

LIST OF ROUTES IN JAMMU AND KASHMIR.—A Tabulated description of over Eighty Routes shewing distance marches, natural characteristics, altitudes, nature of supplies, transport, etc. By Major-General MARQUIS DE BOURBEL.

THACKER, SPINK AND CO., CALCUTTA.

KASHMIR EN FAMILLE. HINTS ON TRAVELLING IN KASHMIR FOR Married People and Children. By Major E. A. BURROWS. R.A.

HANDBOOK FOR VISITORS TO AGRA AND ITS NEIGHBOURHOOD By H. G. KEENE, C.S. Fifth Edition, Revised. Maps, Plans, &c. Fcap. 8vo, cloth. Rs. 2-8.

A HANDBOOK FOR VISITORS TO DELHI AND ITS NEIGHBOUR-HOOD. By H. G. KEENE, C.S. Third Edition. Maps. Fcap. 8vo, cloth. Rs. 2-8.

A HANDBOOK FOR VISITORS TO ALLAHABAD, CAWNPORE, LUCKNOW AND BENARES. By H. G. KEENE. Second Edition, Revised. [*In preparation*

HILLS BEYOND SIMLA. — THREE MONTHS' TOUR FROM SIMLA through Bussahir, Kunowar, and Spiti to Lahoul. ("In the Footsteps of the Few.") By Mrs. J. C. MURRAY-AYNSLEY. Crown 8vo, cloth. Rs. 8.

THACKER'S GUIDE TO DARJEELING.—With two Maps. Fcap. 8vo sewed. Rs. 2.

THE 4-ANNA RAILWAY GUIDE.—With Maps. Published Monthly. As. 4.

THACKER'S GUIDE TO CALCUTTA. By EDMUND MITCHELL. Fcap. 8vo, sewed. Re. 1.

CALCUTTA TO LIVERPOOL, BY CHINA, JAPAN, AND AMERICA, IN 1877. By Lieut.-General Sir HENRY NORMAN. Second Edition. Fcap. 8vo, cloth. Rs. 2-8.

GUIDE TO MASURI, LANDAUR, DEHRA DUN, AND THE HILLS NORTH OF DEHRA; including Routes to the Snows and other places of note; with Chapter on Garhwa (Tehri), Hardwar, Rurki, and Chakrata. By JOHN NORTHAM. Rs. 2-8.

THE SPORTSMAN'S MANUAL.—IN QUEST OF GAME IN KULLU, Lahoul, and Ladak to the Tso Morari Lake, with Notes on Shooting in Spiti, Bara Bagahal, Chamba, and Kashmir, and a detailed description of Sport in more than 100 Nalas. With nine Maps. By Lt.-Col. R. H. TYACKE, late H. M.'s 98th & 34th Regts. Foap. 8vo, cloth. Rs. 8-8.

FROM THE CITY OF PALACES TO ULTIMA THULE.—With a Map of Iceland, Icelandic Vocabulary, Money Tables, &c. By H. K. GORDON. Crown 8vo, sewed. Re. 1.

THACKER'S INDIAN DIRECTORIES AND MAPS.

MAP OF THE CIVIL DIVISIONS OF INDIA.—Including Governments Divisions and Districts, Political Agencies, and Native States; also the Cities and Towns with 10,000 Inhabitants and upwards. Coloured. 20 in. x 36 in. Folded, Re. 1. On linen, Rs. 2.

THACKER, SPINK AND CO., CALCUTTA.

CALCUTTA.—PLANS OF THE OFFICIAL AND BUSINESS PORTION, with houses numbered, and Index of Government Offices and Houses of Business on the Map. Plan of the Residence portion of Calcutta with houses numbered so that their position may easily be found. Two maps in pocket case. The maps are on a large scale. Rs. 1.

1895.—THACKER'S INDIAN DIRECTORY.—Official, Legal, Educational, Professional, and Commercial Directories of the whole of India, General Information; Holidays, &c.; Stamp Duties, Customs Tariff, Tonnage Schedules; Post Offices in India, forming a Gazetteer; List of Governors-General and Administrators of India from beginning of British Rule; Orders of the Star of India, Indian Empire, &c.; Warrant of Precedence. Table of Salutes, &c.; The Civil Service of India; An Army List of the Three Presidencies; A Railway Directory; A Newspaper and Periodical Directory; A Conveyance Directory; Tea, Indigo, Silk, and Coffee Concerns; List of Clubs in India; Alphabetical List of Residents. In thick Royal Octavo. With a Railway Map of India. A Map of the Official and Business portion of Calcutta and a Map of the European Residence portion of Calcutta. Rs. 20.

A COMPLETE LIST OF INDIAN AND CEYLON TEA GARDENS, Indigo Concerns, Silk Filatures, Sugar Factories, Cinchona Concerns, Coffee Estates Cotton and Jute Mills, Collieries, Mines, etc. With their Capital, Directors, Proprietors, Agents, Managers, Assistants, &c., and their Factory Marks by which the Chests may be identified in the Market. [1895.] Rs. 3.

THACKER'S MAP OF INDIA, WITH INSET PHYSICAL MAPS, SKETCH PLANS of Calcutta, Bombay, and Madras. Edited by J. G. BARTHOLOMEW. Corrected to present date. With Railways, Political Changes, and an Index of 10,000 Names, being every place mentioned in "Hunter's Imperial Gazetteer."   •

" An excellent map."—*Glasgow Herald.*
" This is a really splendid map of India, produced with the greatest skill and care."—*Army and Navy Gazette.*
" For compactness and completeness of information few works surpassing or approaching it have been seen in cartography."—*Scotsman.*

NATURAL HISTORY. BOTANY. ETC.

THE FUTURE OF THE DATE PALM IN INDIA (PHŒNIX DACTYLIPTERA). By E. BONAVIA, M.D., Brigade-Surgeon, Indian Medical Department. Crown 8vo. cloth. Rs. 2-8.

GAME, SHORE, AND WATER BIRDS OF INDIA.—BY COL. A. LE MESSURIER, R.E. A *vade mecum* for Sportsmen. With 121 Illustrations. 8vo. Rs. 10.

THACKER, SPINK AND CO., CALCUTTA.

HANDBOOK TO THE FERNS OF INDIA, CEYLON, AND THE MALAY PENINSULA. By Colonel R. H. BEDDOME, author of the "Ferns of British India." With 800 Illustrations by the author. Imp. 16mo. Rs. 10.

"A most valuable work of reference."—*Garden.*

"It is the first special book of portable size and moderate price which has been devoted to Indian Ferns, and is in every way deserving."—*Nature.*

SUPPLEMENT TO THE FERNS OF BRITISH INDIA, CEYLON AND THE MALAY PENINSULA, containing Ferns which have been discovered since the publication of the "Handbook to the Ferns of British India," &c. By Col. R. H. BEDDOME, F.L.S. Crown 8vo, sewed. Rs. 2-12.

GOLD, COPPER, AND LEAD IN CHOTA-NAGPORE.—COMPILED BY W. KING, D.SC., Director of the Geological Survey of India, and T. A. POPE, Deputy Superintendent, Survey of India. With Map showing the Geological Formation and the Areas taken up by the various Prospecting and Mining Companies. Crown 8vo, cloth. Rs. 5.

ON INDIGO MANUFACTURE.—A PRACTICAL AND THEORETICAL GUIDE to the Production of the Dye. With numerous Illustrative Experiments. By J. BRIDGES LEE, M.A., F.G.S. Crown 8vo, cloth. Rs. 4.

"The book is thoroughly practical, and is as free from technicalities as such a work can well be, and it gives as much information as could well be imparted in so small a compass."—*Indian Daily News.*

"Instructive and useful alike to planter and proprietor . . . A very clear and undoubtedly valuable treatise for the use of practical planters, and one which every planter would do well to have always at hand during his manufacturing season. For the rest, a planter has only to open the book for it to commend itself to him."—*Pioneer.*

MANUAL OF AGRICULTURE FOR INDIA.—BY LIEUT. FREDERICK POGSON. Illustrated. Crown 8vo, cloth, gilt. Rs. 5.

THE CULTURE AND MANUFACTURE OF INDIGO.—With a Description of a Planter's Life and Resources. By WALTER MACLAGAN REID. Crown 8vo. With 19 Full-page Illustrations. Rs. 5.

"It is proposed in the following Sketches of Indigo Life in Tirhoot and Lower Bengal to give those who have never witnessed the manufacture of Indigo, or seen an Indigo Factory in this country, an idea of how the finished marketable article is produced: together with other phases and incidents of an Indigo Planter's life, such as may be interesting and amusing to friends at home."—*Introduction.*

ROXBURGH'S FLORA INDICA; OR, DESCRIPTION OF INDIAN PLANTS. Reprinted *litteratim* from Cary's Edition. 8vo, cloth. Rs. 5.

THACKER, SPINK AND CO.,

A NATURAL HISTORY OF THE MAMMALIA OF INDIA, BURMAH
AND CEYLON. By R. A. STERNDALE. F.R.G.S., F.Z.S., &c., author
"Seonee." "The Denizens of the Jungle." With 170 Illustrations by
the author and others. Imp. 16mo. Rs. 0.

"The very model of what a popular natural history should be,"—*Knowledge.*

"The book will, no doubt, be specially useful to the sportsman, and, indeed,
has been extended so as to include all territories likely to be reached by the
sportsman from India."—*The Times.*

A TEA PLANTER'S LIFE IN ASSAM.—By GEORGE M. BARKER.
With 75 Illustrations by the author. Crown 8vo. Rs. 6-8.

"Mr. Barker has supplied us with a very good and readable description
accompanied by numerous illustrations drawn by himself. What may be called
the business parts of the book are of most value."—*Contemporary Review.*

"Cheery, well-written little book."—*Graphic.*

"A very interesting and amusing book, artistically illustrated from sketches
drawn by the author."—*Mark Lane Express.*

A TEXT-BOOK OF INDIAN BOTANY: MORPHOLOGICAL, PHYSIOLOGI-
CAL, and SYSTEMATIC. By W. H. GURGG, B.M.S., Lecturer on Botany at
the Hugli Government College. Profusely Illustrated. Crown 8vo. Rs. 5.

THE INLAND EMIGRATION ACT, AS AMENDED BY ACT VII OF 1893
The Health Act: Sanitation of Emigrants; The Artificer's Act; Land
Rules of Assam, etc. Crown 8vo, cloth. Rs. 2.

ENGINEERING, SURVEYING, ETC.

STATISTICS OF HYDRAULIC WORKS, AND HYDROLOGY OF ENG-
LAND, CANADA, EGYPT, AND INDIA. Collected and reduced by LOWIS
D'A. JACKSON. C.K. Royal 8vo. Rs. 10.

PERMANENT WAY POCKET-BOOK.—CONTAINING COMPLETE FOR-
MULÆ for Laying Points, Crossings, Cross-over Roads, Through Roads,
Diversions, Curves, etc., suitable for any Gauge. With Illustrations. By
T. W. JONKS. Pocket-Book Form, cloth. Rs. 8.

A HANDBOOK OF PRACTICAL SURVEYING FOR INDIA.—Illustrated with Plans, Diagrams etc. Fourth Edition, Revised. By F. W.
KELLY, late of the Indian Survey. With 24 Plates. 8vo. Rs. 8.

PROJECTION OF MAPS. By R. SINCLAIR. With Diagrams. Foolscap
8vo.

SPINK AND CO., CALCUTTA.

IRRIGATED INDIA.—AN AUSTRALIAN VIEW OF INDIA AND CEYLON, their Irrigation and Agriculture. By the Hon. ALFRED DEAKIN, M.L.A., formerly Chief Secretary and Minister of Water-Supply of Victoria, Australia. With a Map. 8vo, cloth. Rs. 7-8.

CONTENTS:—Introduction—India and Australia—The British in India—The Native Population—Physical and Political Divisions—Ceylon—Madras—Lower Bengal—Bombay—The Independent States—The North-West Provinces and the Punjab—The Agriculture of India—Indian Wheat and Australian Trade —Irrigation Generally—The Kaveri Scheme—Ekruk and Khadahvasla— Powai, Vehar and Tansa—The Ganges Canal System—The Bari Doab Canal —The Sirhind Canal—Indian Irrigation.

APPENDICES:—Irrigation in Ceylon—Irrigation in Madras—Madras Company's Canal—Irrigation in Bombay—Irrigation in Lower Bengal—Irrigation in the North-West Provinces—Irrigation in the Punjab.

" I think that I may again with profit refer to Mr. Deakin's Book on Irrigated India, the perusal of which I am glad to have this opportunity of recommending to the attention of those who are interested in the welfare of this country."
—*C. W. Odling,* M.E., *in a Lecture on Irrigation Canals, delivered at Sibpur.*

" He approaches Indian problems with an Australian freshness of view and frankness of comment that are often singularly suggestive."—*Times.*

" Contains a masterly account of the great gift of the English to India— the irrigation works."—*Manchester Guardian.*

"It is the work of an observer of no ordinary capacity and fitness for the work of observing and describing."—*Scotsman.*

AN EXPLANATION OF QUADRUPLEX TELEGRAPHY. — With 2 Diagrams. By BEN. J. STOW, Telegraph Master. Fcap. 4to, Rs. 2.

AN EXPLANATION ON DUPLEX, QUADRUPLEX, OPEN AND TRANS-LATION WORKING AND OTHER CIRCUITS.—Testing of Currents, Batteries, Instruments, Earths, and Line, with the Tangent Galvanometer. With 12 Plates. By E. H. NELTHROPP, Telegraph Master. Crown 8vo, sewed. Rs. 2.

MANUAL OF SURVEYING FOR INDIA.—DETAILING THE MODE OF operations on the Trigonometrical, Topographical, and Revenue Surveys of India. By Col. H. L. THUILLIER and Lieut.-Col. H. SMYTH. Third Edition, Revised and Enlarged. Royal 8vo, cloth. Rs. 12.

COLEBROOKE'S TRANSLATION OF THE LILAVATI.—With Notes. By HARAN CHANDRA BANERJI, M.A., B.L. 8vo, cloth. Rs. 4.

This edition includes the Text in Sanskrit. The Lilavati is a standard work on Hindu mathematics written by Bháskaráchárya, a celebrated mathematician of the twelfth century.

RAILWAY CURVES.—Practical Hints on Setting out Curves, with a Table of Tangents for a 1" Curve for all angles from 2° to 135° increasing by minutes: and other useful Tables. With a Working Plan and Section of Two miles of Railway. By A. G. WATSON, Assistant Engineer. Rs. 4.

THACKER, SPINK· AND CO., CALCUTTA.

FIRE INSURANCE IN INDIA.—A short account of the Principles and Practice of Fire Insurance, Settlement of Losses, Extinction and prevention of Fire, &c. By BREMAWALIAH. Crown 8vo. Sewed. Re. 1-8.

A HANDBOOK OF PHOTOGRAPHY.—WRITTEN ESPECIALLY FOR INDIA. By GEORGE EWING, Honorary Treasurer of the Photographic Society of India. [*In the Press.*

THE PHOTOGRAPHER'S POCKET-BOOK.—A Compilation of all Information regarding Photography in a small handy form. [*In the Press.*

THE JOURNAL OF THE PHOTOGRAPHIC SOCIETY.—Published Monthly. With Illustrations. Subscription Rs. 5 yearly.

MILITARY WORKS.

THE RECONNOITRER'S GUIDE AND FIELD BOOK.—ADAPTED FOR INDIA. By Major M. J. KING-HARMAN, B.S.C. Second Edition, Revised and Enlarged. In roan. Rs. 4.

It contains all that is required for the guidance of the Military Reconnoitrer in India: it can be used as an ordinary Pocket Note Book, or as a Field Message Book ; the pages are ruled as a Field Book, and in sections, for written description or sketch.

The book has been highly approved by Lord Roberts, who regards it as a most valuable and practical composition.

"To Officers serving in India the Guide will be invaluable."—*Broad Arrow.*

"It appears to contain all that is absolutely required by the Military Reconnoitrer in India, and will thus dispense with many bulky works. In fact it contains just what is wanted and nothing not likely to be wanted."—*Naval and Military Gazette.*

"It has been found invaluable to many a Staff Officer and Commandant of a Regiment. as well as of the greatest possible assistance to officers studying for the Garrison Course Examination. The book will go into the breast-pocket of a regulation khaki jacket, and can therefore always fulfil the office of a *vade mecum.*"—*Madras Mail.*

INDIAN ARTICLES OF WAR, ARRANGED FOR EASY REFERENCE. By Major F. W. KITCHENER. [*In the Press.*

THE QUARTERMASTER'S ALMANAC.—A DIARY OF THE DUTIES, with other information. By Lieut. HARRINGTON BUSH. 8vo. Re. 1-8.

THACKER, SPINK AND CO., CALCUTTA.

LETTERS ON TACTICS AND ORGANIZATION.—By CAPT. F. N. MAUDE, R.E. (Papers reprinted from the *Pioneer* and *Civil and Military Gazette.*) Crown 8vo, cloth. Rs. 5.

"The author displays considerable knowledge of the subjects with which he deals, and has evidently thought much on them. His views are broad and advanced."—"Every soldier should read this book."—*Athenæum.*

"On the whole, Captain Maude may be most warmly congratulated upon the production of a book, of which, disagreeing as we do with some of his conclusions, we are glad to speak, as it deserves, in terms of the most unstinted and ungrudging praise."—*Whitehall Review.*

THE INVASION AND DEFENCE OF ENGLAND.—By CAPT. F. N. MAUDE R.E. Crown 8vo, cloth. Rs. 1-8.

"This little book only deals with the case of possible invasion by France, but it is one of the best we have read on the subject, and will well repay perusal."—*Allen's Indian Mail.*

"His little book is a useful and interesting contribution to the invasion of England question; it contains a good deal of information, and, without being written in an alarmist style, exposes very clearly the danger in which England stands."—*Englishman.*

"The lay reader will welcome as an able, thoughtful, and original contribution to a topic of unsurpassable importance."—*Home News.*

"The book is ably written, and is full of suggestive matter of the highest importance to the security of the country."—*Glasgow Herald.*

THE SEPOY OFFICER'S MANUAL. By CAPT. E. G. BARROW. Third Edition, Entirely Re-written, and brought up to date. By CAPT. E. H. BINGLEY, 7th Bengal Infantry. 12mo, cloth, Rs. 2-8.

"It seems to contain almost everything required in one of the modern type of Civilian Soldiers . . . . In the most interesting part of the book is an account of the composition of the Bengal Army with descriptive note on the Brahmans, Rajputs, Sikhs, Goorkhas, Pathans and other races."—*Englishman.*

"A vast amount of technical and historical data of which no Anglo-Indian Officer should be ignorant."—*Broad Arrow.*

"The notes are brief and well digested, and contain all that it is necessary for a candidate to know."—*Army and Navy Gazette.*

THACKER, SPINK AND CO., CALCUTTA.

THE INDIAN ARTICLES OF WAR. ANNOTATED. BY CAPT. H. S.
HUDSON, late 27th Madras Infantry. Third Edition. Revised in
accordance with the amended Indian Articles of War. By CAPT. C. E.
POYNDER. Crown 8vo, cloth.

" Likely to be useful to Examiners."—*Army and Navy Gazette.*
" Complete, intelligible, and attractive."—*Englishman.*
" Extremely useful to those who have to deal with cases rising under the
Indian Articles of War."—*Broad Arrow.*

THE INDIAN MESSAGE BOOK.—INTERLEAVED FOR KEEPING COPIES.
With 12 Authorised Pattern Envelopes. Each Re. 1-4.

NOTES ON THE COURSE OF GARRISON INSTRUCTION, TACTICS,
Topography, Fortifications, condensed from the Text-Books, with expla-
nations and additional matter. With Diagrams. By Major E. LLOYD,
Garrison Instructor. Crown 8vo, cloth. Rs. 2-8.

LECTURES DELIVERED TO TRANSPORT CLASSES.—A complete
Epitome of Transport Duties and Veterinary for use in Classes and for
Ready Reference in the Field. By a Deputy Assistant Commissary-
General. [*In the Press.*

HINDUSTANI, PERSIAN, ETC.

GLOSSARY OF MEDICAL AND MEDICO-LEGAL TERMS, including
those most frequently met with in the Law Courts. By R. F. HUTCHI-
SON, M.D., Surgeon-Major. Second Edition. Fcap. 8vo, cloth. Rs. 2.

HIDAYAT AL HUKUMA.—A GUIDE TO MEDICAL OFFICERS AND SUB-
ORDINATES of the Indian Service. English and Hindustani. By GEO. S.
RANKING, M.D., Surgeon-Major. 18mo, sewed. Re. 1-4.

THE DIVAN-I-HAFIZ.—THE DIVAN WRITTEN IN THE FOURTEENTH
CENTURY by Khwaja-Shame-ud-din Mohammad-i-Hafiz-i-Shirazi, translated for the first time out of the Persian into English Prose, with Critical and Explanatory remarks, with an Introductory Preface, a Note on
Sufi'ism, and Life of the author. By Lieut.-Col. H. WILBERFORCE
CLARKE, author of " The Persian Manual," translator of " The Bustan-
i-Sa'di," " The Sekandar Namah-i-Nizami," etc. 2 vols. 4to. Rs. 25,

---

THACKER, SPINK AND CO., CALCUTTA.

THE 'AWARIFU-L-MA'ARIF.—WRITTEN IN THE THIRTEENTH CENTURY by Shaikh Shahab-ud-din—'Umar bin Muhammad-i-Sahrwardi translated (out of the Arabic into Persian) by Mamud bin 'Ali al Kashani, Companion in Sufi'ism to the Divan-i-Khwaja Hafiz; translated for the first time (out of the Persian into English) by Lieut.-Col. H. WILBERFORCE CLARKE. 4to. Rs. 18.

HISTORY OF THE SIKHS: OR, TRANSLATION OF THE SIKKHAN DE RAT DI VIKHIA, as laid down for the Examination in Panjabi, &c., together with a short Gurmukhi Grammar. By Lt.-Col. MAJOR HENRY COURT. Royal 8vo, cloth. Rs. 8.

THE RUSSIAN CONVERSATION GRAMMAR.—BY ALEX. KINLOCH, late Interpreter to H. B. M. Consulate and British Consul in the Russian Law Courts; Instructor for Official Examinations. Crown 8vo, cloth. Rs. 6-8.

This work is constructed on the excellent system of Otto in his "German Conversation Grammar," with illustrations accompanying every rule, in the form of usual phrases and idioms, thus leading the student by easy but rapid gradations to a colloquial attainment of the language.

UTTARA RAMA CHARITA.—A SANSKRIT DRAMA. By BHAVABHUTI Translated into English Prose by C. H. TAWNEY, M.A. Second Edition. Adapted to Pundit I. C. VIDYASAGARA's Edition of the Text.

ROMANISED URDU GRAMMAR.—Compiled upon new lines, giving the gist of large Grammars in a small compass. By Rev. G. SMALL, M.A., late Missionary in Bengal. Crown 8vo. Rs. 5.

AN ANGLO-INDIAN MEDICAL MANUAL AND VOCABULARY. By Rev. G. SMALL, M.A. Crown 8vo, cloth, limp. Rs. 5.

The object of which is to furnish to medical practitioners, male and female, as well as the general public, a help in the practice of medicine in India. The author has received the valuable co-operation of Surgeon-General R. S. Francis, late Medical College, Calcutta.

THACKER, SPINK AND CO., CALCUTTA.

A GUIDE TO HINDUSTANI (TALIM-I-ZABAN-I-URDU). Specially designed for the use of students and men serving in India, By Surgeon-Major GEO. S. RANKING, Offg, Secretary to the Board of Examiners, Fort William. Third Edition. 8vo, cloth.   [*In the Press.*

Printed throughout in Persian character. With *fac-simile* MS. Exercises, Petitions, &c.

"The work on the whole, we believe, will meet a want. . . . . . It contains an excellent list of technical military terms and idioms, and will prove especially serviceable to any one who has to act as an interpreter at courts-martial and cognate enquiries."—*Civil and Military Gazette.*

"There can be no question as to the practical utility of the book."—*Pioneer.*

"Surgeon-Major Ranking has undoubtedly rendered good service to the many military men for whom knowledge of Hindustani is essential."—*Athenæum.*

"Has the merit of conciseness and portability, and the selections at the end, of the historical and colloquial style, are well chosen."—*Saturday Review.*

"A well-conceived book, and has much useful matter in it. The sentences are very good, practical and idiomatic."—*Homeward Mail.*

"Supplies a want long felt, by none more than by young Medical Officers of the Army of India. We think the work admirably adapted for its purpose."—*British Medical Journal.*

MALAVIKAGNIMITRA.—A SANSKRIT PLAY BY KALIDASA. Literally translated into English Prose by C. H. TAWNEY, M.A., Principal, Presidency College, Calcutta. Second Edition. Crown 8vo, Re. 1-8.

TWO CENTURIES OF BHARTRIHARI.—TRANSLATED INTO ENGLISH VERSE by C. H. TAWNEY, M.A. Fcap. 8vo, cloth. Rs. 2.

HINDUSTANI AS IT OUGHT TO BE SPOKEN.—BY J. TWEEDIE, Bengal Civil Service. Second Edition. Crown 8vo, pp. xvi, 850, cloth. Rs. 4-8.

SUPPLEMENT containing Key to the Exercises and Translation of the Reader with Notes. Rs. 2.

The work has been thoroughly Revised and partly Re-Written, and much additional matter added. The VOCABULARIES have been improved, and all words used in the book have been embodied in the GLOSSARIES, ENGLISH-HINDUSTANI—HINDUSTANI-ENGLISH. A READER is also given, and a GENERAL INDEX to the whole book.

"The Young Civilian or Officer, reading for his Examination, could not do better than master this Revised Edition from cover to cover."—*I. Daily News.*

"The book is divided into twelve easy lessons, and there is nothing to prevent the most khansamah-worried *mem-saheb* from mastering one of these a day. At the end of a fortnight she will have acquired a small useful vocabulary, and should be quite certain how to use the words she knows."—*Englishman.*

THACKER, SPINK AND CO., CALCUTTA.

BOOK-KEEPING AND OFFICE MANUALS.

A GUIDE TO BOOK-KEEPING.—BY SINGLE, MIXED AND DOUBLE ENTRIES. Commercial Accounts of the most intricate nature fully illustrated by Examples and Annotations; Answers to Examination Questions Book-Keeping, for Promotion to Assistant Examiner (1st grade) and to Accountant (2nd grade), from 1880 to 1891. By S. GEORGE, late Chief Accountant, P. W. D., Bengal. Demy 8vo, cloth. Rs. 2-8.

PHONOGRAPHY IN BENGALI.—BY DWIJENDRA NATH SHINGHAW, Professor of Phonography in Calcutta. Being a Handbook for the study of Shorthand on the principle of Pitman's System. 12mo. As. 8. With a Key. 12mo. As. 4 extra.

THE INDIAN SERVICE MANUAL; OR, GUIDE TO THE SEVERAL DEPARTMENTS of the Government of India, containing the Rules for Admission, Notes on the working of each Department, &c. By C. R. HANDLESS, author of "The Clerk's Manual."

THE GOVERNMENT OFFICE MANUAL.—A GUIDE TO THE DUTIES, Privileges and Responsibilities of the Government Service in all Grades. By CHARLES HANDLESS. Crown 8vo. Rs. 2.

SPENS' THE INDIAN READY RECKONER.—CONTAINING TABLES for ascertaining the value of any number of articles, &c., from three pies to five rupees; also Tables of Wages from four annas to twenty-five rupees. By Captain A. T. SPENS. Re. 1-8.

THE INDIAN LETTER-WRITER.—CONTAINING AN INTRODUCTION on Letter Writing, with numerous Examples in the various styles of Correspondence. By H. ANDERSON. Crown 8vo, cloth. Re. 1.

THE CLERK'S MANUAL.—A COMPLETE GUIDE TO GENERAL OFFICE ROUTINE (Government and Business). By CHARLES R. HANDLESS. Second Edition, Revised. 12mo, boards. Rs. 2.

INDIAN WAGES TABLES.—CALCULATED FOR MONTHS OF 28 TO 31 working days at rates from 2 to 18 rupees per month, giving the calculation at Sight for 1 to 1,000 days from 2 to 8½ rupees per month and to 20,000 days by one addition: and for 1 to 800 days from 9 to 18 rupees per month. Also 3 Tables of Sirdaree for those who require them. By G. G. PLAYFAIR, Secretary of the Lebong Tea Co., Ld., and formerly one of the Brahmaputra Tea Co., Ld. [*In the Press.*

THACKER, SPINK AND CO., CALCUTTA.

EDUCATIONAL BOOKS.

HINTS ON THE STUDY OF ENGLISH.—By F. J. ROWE, M.A., and W. T. WEBB, M.A., Professors of English Literature, Presidency College, Calcutta. New Edition. With an additional chapter on the Structure and Analysis of Sentences, and Exercises on the correction of mistakes commonly made by Students. Crown 8vo, cloth. Rs. 2-8.

AN ELEMENTARY ENGLISH GRAMMAR FOR SCHOOLS IN INDIA.—Containing numerous Exercises in Idiom. By F. J. ROWE, M.A., and W. T. WEBB, M.A., authors of "Hints on the Study of English." Fcap. 8vo, cloth. Re. 1.

A COMPANION READER TO "HINTS ON THE STUDY OF ENGLISH." (Eighteenth Thousand.) Demy 8vo. Re. 1-4.

A KEY TO THE COMPANION READER TO "HINTS ON THE STUDY OF ENGLISH." With an Appendix, containing Test Examination Questions. By F. J. ROWE, M.A. Fcap. 8vo. Rs. 2.

ENTRANCE TEST EXAMINATION QUESTIONS AND ANSWERS in English, being the Questions appended to "Hints on the Study of English," with their Answers, together with Fifty Supplementary Questions and Answers. By W. T. WEBB. M.A. 12mo, sewed. Re. 1.

PRINCIPAL EVENTS IN INDIAN AND BRITISH HISTORY.—With their Dates in Suggestive Sentences. In Two Parts. By Miss ADAMS, La Martinière College for Girls, Calcutta. Second Edition. Demy 8vo, boards. Re. 1.

ELEMENTARY STATICS AND DYNAMICS.—By W. N. BOUTFLOWER, B.A., late Scholar of St. John's College, Cambridge, and Professor of Mathematics, Muir Central College, Allahabad. Second Edition. Crown 8vo. Rs. 3-8.

THE STUDENT'S HANDBOOK TO HAMILTON AND MILL.—By W. BELL, M.A., Professor of Philosophy and Logic, Government College, Lahore. 8vo, boards. Rs. 2.

ELEMENTARY HYDROSTATICS.—WITH NUMEROUS EXAMPLES AND UNIVERSITY PAPERS. By S. B. MUKERJEE, M.A., B.L., Assistant Professor, Government College, Lahore. 12mo, cloth. Re. 1-8.

PROJECTION OF MAPS. By R. SINCLAIR. With Diagrams. Foolscap. 8vo.

THACKER, SPINK AND CO., CALCUTTA.

ENGLISH SELECTIONS APPOINTED BY THE SYNDICATE OF THE CAL-CUTTA UNIVERSITY for the Entrance Examination. Crown 8vo, cloth. Re. 1-8.

WEBB'S KEY TO THE ENTRANCE COURSE.—1894 and 1895. *Each* Rs. 2.

THE LAWS OF WEALTH.—A PRIMER ON POLITICAL ECONOMY FOR THE MIDDLE CLASSES IN INDIA. By HORACE BELL, C.K. Seventh Thousand. Fcap. 8vo. As. 8.

THE INDIAN LETTER-WRITER.—CONTAINING AN INTRODUCTION ON LETTER WRITING, with numerous Examples in the various styles of Correspondence. By H. ANDERSON. Crown 8vo, cloth. Re. 1.

A CATECHISM ON THE RUDIMENTS OF MUSIC.—SIMPLIFIED FOR BEGINNERS. By I. LITTLEPAGE. 12mo, sewed. Re. 1.

CALCUTTA UNIVERSITY CALENDAR FOR THE YEAR 1894.—Containing Acts, Bye-Laws, Regulations, The University Rules for Examination, Text-Book Endowments, Affiliated Institutions, List of Graduates and Under-Graduates, Examination Papers, 1893. Cloth. Rs. 6. CALENDAR for previous years. *Each* Rs. 5.

CALCUTTA UNIVERSITY CALENDAR.—THE EXAMINATION PAPERS, 1890 and 1891. Cloth. *Each* Re. 1-8.

FIFTY GRADUATED PAPERS IN ARITHMETIC, ALGEBRA, AND GEOMETRY for the use of Students preparing for the Entrance Examinations of the Indian Universities. With Hints on Methods of Shortening Work and on the Writing of Examination Papers. By W. H. WOOD, B.A., F.C.S., Principal, La Martinière College. Re. 1-8.

THE PRINCIPLES OF HEAT.—FOR THE F. A. EXAMINATION of the Calcutta University. By LEONARD HALL, M.A. Crown 8vo. As. 8.

ANALYSIS OF REID'S ENQUIRY INTO THE HUMAN MIND.—With Copious Notes. By W. C. FINK. Second Edition. Re. 1-12.

THE ENGLISH PEOPLE AND THEIR LANGUAGE.—Translated from the German of Loth by C. H. TAWNEY, M.A., Professor in the Presidency College, Calcutta. Stitched. As. 8.

TALES FROM INDIAN HISTORY.—BEING THE ANNALS OF INDIA retold in Narratives. By J. TALBOYS WHEELER. Crown 8vo, cloth. School Edition. Re. 1-8.

A NOTE ON THE DEVANAGARI ALPHABET FOR BENGALI STUDENTS. By GURU DAS BANERJEE, M.A., D.L. Crown 8vo. As. 4.

THACKER, SPINK AND CO., CALCUTTA.

THE GOVERNMENT OF INDIA.—A PRIMER FOR INDIAN SCHOOLS.
By HORACE BELL, C.E. Third Edition. Fcap. 8vo, sewed, As. 8; in
cloth, Re. 1.

Translated into Bengali. By J. N. BHATTACHARJEE. 8vo. As. 12.

AN INQUIRY INTO THE HUMAN MIND ON THE PRINCIPLES OF
COMMON SENSE. By THOMAS REID, D.D. 8vo, cloth. Re. 1-4.

A TEXT-BOOK OF INDIAN BOTANY: MORPHOLOGICAL, PHYSIOLOGI-
CAL, and SYSTEMATIC. By W. H. GREGG, B.M.S., Lecturer on Botany at
Hugli Government College. Profusely Illustrated. Crown 8vo. Rs. 5.

A MORAL READING BOOK FROM ENGLISH AND ORIENTAL SOURCES.
By ROPER LETHBRIDGE, C.I.K., M.A. Crown 8vo, cloth. As. 14.

A PRIMER CATECHISM OF SANITATION FOR INDIAN SCHOOLS.—
Founded on Dr. Cunningham's Sanitary Primer. By L. A. STAPLEY.
Second Edition. As. 4.

NOTES ON MILL'S EXAMINATION OF HAMILTON'S PHILOSOPHY.
By THOMAS EDWARDS, F.R.I.S. Fcap. Sewed. Re. 1.

A SHORT HISTORY OF THE ENGLISH LANGUAGE.—By THOMAS
EDWARDS, F.R.I.S. 18mo, Re. 1-4.

LAMB'S TALES FROM SHAKESPEARE.—AN EDITION IN GOOD TYPE.
Cloth. As. 12.

LAND TENURES AND LAND REVENUE.

AZIZUDDIN AHMED.—THE N.-W. PROVINCES LAND REVENUE
ACT. Being Act XIX of 1873 as amended by Acts I and VIII of 1879, XII
of 1881, XIII and XIV of 1882, XX of 1890, and XII of 1891. With Notes,
Government Orders, Board Circulars and Decisions, and Rulings of the
Allahabad High Court. By AZIZUDDIN AHMED, Deputy Collector and
Magistrate. Demy 8vo, cloth. Rs. 8.

BEVERLEY.—THE LAND ACQUISITION ACTS (ACTS I OF 1894 AND
XVIII OF 1885 Mines). With Introduction and Notes. The whole forming
a complete Manual of Law and Practice on the subject of Compensation for
Lands taken for Public Purposes. Applicable to all India. By H.
BEVERLEY, M.A., B.C.S. Third Edition. Cloth gilt. Rs. 6.

THACKER, SPINK AND CO., CALCUTTA.

C

FORSYTH.—REVENUE SALE-LAW OF LOWER BENGAL, comprising Act XI of 1859; Bengal Act VII of 1868; Bengal Act VII of 1880 (Public Demands Recovery Act), and the unrepealed Regulations and the Rules of the Board of Revenue on the subject. With Notes. Edited by WM. E. H. FORSYTH. Demy 8vo, cloth. Rs. 5.

PHILLIPS.—MANUAL OF REVENUE AND COLLECTORATE LAW. With Important Rulings and Annotations. By H. A. D. PHILLIPS, Bengal Civil Service. Crown 8vo, cloth. [1884] Rs. 10.

CONTENTS:—Alluvion and Diluvion, Certificate, Cesses, Road and Public Works, Collectors, Assistant Collectors, Drainage, Embankment, Evidence, Excise, Lakhiraj Grants and Service Tenures, and Land Acquisition, Land Registration. Legal Practitioners, License Tax, Limitation, Opium, Partition, Public Demands Recovery, Putni Sales, Registration, Revenue Sales, Salt, Settlement, Stamps, Survey and Wards.

REYNOLDS.—THE NORTH-WESTERN PROVINCES RENT ACT.— With Notes, &c. By H. W. REYNOLDS, C.S. Demy 8vo. [1886] Rs. 7.

FIELD.—LANDHOLDING, AND THE RELATION OF LANDLORD AND TENANT in various countries of the world. By C. D. FIELD, M.A., LL.D. Second Edition. 8vo, cloth. Rs. 16.

N.B.—This edition contains " The Bengal Tenancy Act, 1885," with Notes and Observations: and an Index to the whole of the Law of Landlord and Tenant in Bengal.

"We may take it that, as regards Indian laws and customs, Mr. Field shows himself to be at once an able and skilled authority. In order, however, to render his work more complete, he has compiled, chiefly from Blue-books and similar public sources, a mass of information having reference to the land-laws of most European countries, of the United States of America, and our Australasian colonies."—*Field.*

GRIMLEY.—MANUAL OF THE REVENUE SALE LAW AND CERTIFICATE PROCEDURE of Lower Bengal, including the Acts on the Subject and Selections from the Rules and Circular Orders of the Board of Revenue. With Notes. By W. H. GRIMLEY, B.A., C.S. 8vo, Rs. 5-8; interleaved, Rs. 6.

PHILLIPS.—THE LAW RELATING TO THE LAND TENURES OF LOWER BENGAL. (Tagore Law Lectures, 1875.) By ARTHUR PHILLIPS. Royal 8vo, cloth. Rs. 10.

THACKER, SPINK AND CO., CALCUTTA.

REGULATIONS OF THE BENGAL CODE.—A SELECTION intended chiefly for the use of Candidates for appointments in the Judicial and Revenue Departments. Royal 8vo, stitched. Rs. 4.

PHILLIPS.—OUR ADMINISTRATION OF INDIA.—BEING A COMPLETE ACCOUNT of the Revenue and Collectorate Administration in all departments, with special reference to the work and duties of a District Officer in Bengal. By H. A. D. PHILLIPS. Rs. 5.

"In eleven chapters Mr. Phillips gives a complete epitome of the civil, in distinction from the criminal, duties of an Indian Collector."—*London Quarterly Review.*

WHISH.—A DISTRICT OFFICE IN NORTHERN INDIA.—With some suggestions on Administration. By C. W. WHISH, B.C.S. Demy 8vo, cloth. Rs. 4.

"Mr. Whish has produced an extremely useful and thoughtful book, which will pave the way for the junior members of his service. It is above all things practical, and sets forth the whole scheme of district duties in a clear and systematic manner."—*Englishman.*

FIELD.—INTRODUCTION TO THE REGULATIONS OF THE BENGAL CODE. By C. D. FIELD, M.A., LL.D. Crown 8vo. Rs. 3.

CONTENTS: (I) The Acquisition of Territorial Sovereignty by the English in the Presidency of Bengal. (II) The Tenure of Land in the Bengal Presidency. (III) The Administration of the Land Revenue. (IV) The Administration of Justice.

MARKBY.—LECTURES ON INDIAN LAW.—BY WILLIAM MARKBY, M.A. Crown 8vo, cloth. Rs. 3.

CONTENTS: (I) Resumption of Lands held Rent-free. (II) The Revenue Sale Land of the Permanently Settled Districts. (III) Shekust Pywust, or Alluvion and Diluvion. (IV-V) The charge of the Person and Property of Minors. (VI) Of the protection afforded to Purchasers and Mortgagees when their title is impeached. Appendix—The Permanent Settlement—Glossary.

HOUSE.—THE N.-W. PROVINCES RENT ACT.—BEING ACT XII o 1881, as amended by subsequent Acts. Edited with Introduction, Commentary and Appendices. By H. F. HOUSE, C.S. 8vo, cloth. Rs. 10.

THACKER, SPINK AND CO., CALCUTTA.

CIVIL LAW.

REGISTRATION ACT AND MANUAL.—By H. HOLMWOOD, I.C.S. Recently Registrar-General of Assurances, Bengal. Royal 8vo.
[*In the Press.*

ALEXANDER.—INDIAN CASE-LAW ON TORTS. By THE LATE R. D. ALEXANDER, C.S. An entirely new Edition, Re-written and Enlarged by R. F. RAMPINI, C.S. 8vo, cloth. Rs. 8.

CHALMERS.—THE NEGOTIABLE INSTRUMENTS ACT, 1881.—Being an Act to define and amend the Law relating to Promissory Notes, Bills of Exchange, and Cheques. Edited by M. D. CHALMERS, M.A., Barrister-at-law, author of "A Digest of the Law of Bills of Exchange," &c.; and editor of Wilson's "Judicature Acts." 8vo, cloth. Rs. 7-8.

COLLETT.—THE LAW OF SPECIFIC RELIEF IN INDIA.—Being a Commentary on Act I of 1877. By CHARLES COLLETT, late of the Madras Civil Service, of Lincoln's Inn, Barrister-at-Law, and formerly a Judge of the High Court at Madras. Second Edition. [*In the Press.*

KELLEHER.—PRINCIPLES OF SPECIFIC PERFORMANCE AND MISTAKE. By J. KELLEHER, C.S. 8vo, cloth. Rs. 8.

"The work is well written, and the rules deduced from the authorities are generally accurately and always clearly expressed. We can therefore recommend the book to all students of English law, not doubting but that they will find it very useful for their purposes."—*Civil and Military Gazette.*

KELLEHER.—MORTGAGE IN THE CIVIL LAW.—Being an outline of the Principles of the Law of Security, followed by the text of the Digest of Justinian, with Translation and Notes; and a translation of the corresponding titles of the Indian Code. By J. KELLEHER, B.C.S., author of "Possession in the Civil Law." Royal 8vo. Rs. 10.

KELLEHER.—POSSESSION IN THE CIVIL LAW.—Abridged from the Treatise of Von Savigny, to which is added the Text of the Title on Possession from the Digest. By J. KELLEHER, C.S. 8vo, cloth. Rs. 8.

SUTHERLAND.—THE INDIAN CONTRACT ACT (IX OF 1872) AND THE SPECIFIC RELIEFS ACT (I OF 1877). With a Full Commentary. By D. SUTHERLAND. Second Edition. Royal 8vo, cloth. Rs. 5.

CASPERSZ.—THE LAW OF ESTOPPEL IN INDIA.—Part I, Estoppel by Representation. Part II, Estoppel by Judgment. Being Tagore Law Lectures, 1893. By A. CASPERSZ, Bar.-at-Law. Royal 8vo, cloth. Rs. 12.

THE INDIAN INSOLVENCY ACT.—BEING A REPRINT OF THE LAW as to Insolvent Debtors in India, 11 and 12 Vict. Cap. 21 (June 1848). Royal 8vo, sewed. (Uniform with Acts of the Legislative Council.) Rs. 1-8.

THACKER, SPINK AND CO., CALCUTTA.

THE LAW OF FRAUD, MISREPRESENTATION AND MISTAKE IN
British India. By Sir Frederick Pollook, Bart., Barrister-at-Law,
Professor of Jurisprudence, Oxford. Being the Tagore Lectures, 1894.
Royal 8vo, cloth gilt. Rs. 10.

A TREATISE ON THE LAW OF RES JUDICATA: Including the
Doctrines of Jurisdiction, Bar by Suit and Lis Pendens. By Hukm
Chand, m.a. In Royal 8vo, cloth. Rs. 24.

RIVAZ.—THE INDIAN LIMITATION ACT (Act XV of 1877) as amend-
ED to date. With Notes. By the Hon'ble H. T. Rivaz, Barrister-at-Law,
Judge of the High Court of the Punjab. Fourth Edition. Royal 8vo, cloth.
Rs. 10.

SUCCESSION, ADMINISTRATION, etc.

FORSYTH.—THE PROBATE AND ADMINISTRATION ACT.—Being
Act V of 1881. With Notes. By W. E. H. Forsyth. Edited, with
Index, by F. J. Collinson. Demy 8vo, cloth. Rs. 5.

HENDERSON.—THE LAW OF INTESTATE AND TESTAMENTARY
Succession in India; including the Indian Succession Act (X of 1865),
with a Commentary; and the Parsee Succession Act (XXI of 1865), the
Hindu Wills Act (XXI of 1870), the Probate and Administration Act, &c.
With Notes and Cross References. By Gilbert S. Henderson, m.a.,
Barrister-at-Law, and Advocate of the High Court at Calcutta.

HENDERSON.—THE LAW OF TESTAMENTARY DEVISE.—As administered in India, or the Law relating to Wills in India. With an Appendix, containing:—The Indian Succession Act (X of 1865), the Hindu
Wills Act (XXI of 1870), the Probate and Administration Act (V of 1881
with all amendments, the Probate Administration Act (VI of 1889), and the
Certificate of Succession Act (VII of 1889). By G. S. Henderson, m.a.,
Barrister-at-Law. (Tagore Law Lectures, 1887.) Royal 8vo, cloth. Rs. 16.

CIVIL PROCEDURE, SMALL CAUSE COURT, etc.

BROUGHTON.—THE CODE OF CIVIL PROCEDURE.—Being Act X
of 1877. With Notes and Appendix. By the Hon'ble L. P. Delves
Broughton. assisted by W. F. Agnew and G. S. Henderson. Royal
8vo. cloth. Reduced to Rs. 7.

O'KINEALY.—THE CODE OF CIVIL PROCEDURE (Act XIV of 1882).
With Notes, Appendices, &c. By the Hon'ble J. O'Kinealy. Fourth
Edition. Royal 8vo. Rs. 16.

THACKER, SPINK AND CO., CALCUTTA.

MACEWEN.—THE PRACTICE OF THE PRESIDENCY COURT OF SMALL CAUSES OF CALCUTTA, under the Presidency Small Cause Courts Act (XV of 1882). With Notes and an Appendix. By R. S. T. Mac-Ewen, of Lincoln's Inn, Barrister-at-Law, one of the Judges of the Presidency Court of Small Causes of Calcutta. Thick 8vo. Rs. 10.

POCKET CODE OF CIVIL LAW.—CONTAINING THE CIVIL PROCEDURE CODE (Act XIV of 1882), The Court-Fees Act (VII of 1870), The Evidence Act (I of 1872), The Specific Reliefs Act (I of 1877), The Registration Act (III of 1877), The Limitation Act (XV of 1877), The Stamp Act (I of 1879). With Supplement containing the Amending Act of 1888, and a General Index. Revised, 1891. Fcap. 8vo, cloth. Rs. 4.

LOCAL SELF-GOVERNMENT.

STERNDALE.—MUNICIPAL WORK IN INDIA.—OR, HINTS ON SANITATION, General Conservancy and Improvement in Municipalities, Towns, and Villages. By R. C. STERNDALE. Crown 8vo, cloth. Rs. 3.

COLLIER.—THE BENGAL LOCAL SELF-GOVERNMENT HANDBOOK. —Being ACT III OF 1885, B. C., and the General Rules framed thereunder. With Notes, Hints regarding Procedure, and References to Leading Cases; an Appendix, containing the principal Acts referred to, &c., &c. By F. R. STANLEY COLLIER, B.C.S. Third Edition, thoroughly revised and brought up to date. Crown 8vo. Rs. 5.

COLLIER.—THE BENGAL MUNICIPAL MANUAL.—BEING B. C. ACT III OF 1884. With Notes and an Appendix containing all the Acts and Rules relating to Municipalities. By F. K. STANLEY COLLIER, C.S. Third Edition. Crown 8vo. Rs. 5.

CRIMINAL LAW.

THE INDIAN PENAL CODE. WITH A COMMENTARY. BY W. R. HAMILTON, Barrister-at-Law. Royal 8vo, cloth. Rs. 16.

COLLETT.—COMMENTS ON THE INDIAN PENAL CODE. By CHARLES COLLETT, Barrister-at-Law. Pvo. Rs. 5.

POCKET PENAL, CRIMINAL PROCEDURE, AND POLICE CODES.— Also the Whipping Act and the Railway Servants' Act, being Acts XLV of 1860 (with Amendments), X of 1882, V of 1861, VI of 1864, and XXXI of 1867. With a General Index. Revised 1892. Fcap. 8vo, cloth. Rs. 4.

AGNEW AND HENDERSON.—THE CODE OF CRIMINAL PROCE-
DURE (ACT X OF 1882), together with Rulings, Circular Orders, Notifica-
tions, &c., of all the High Courts in India, and Notifications and Orders of
the Government of India and the Local Governments. Edited, with
Copious Notes and Full Index, by W. F. AGNEW, Bar.-at-Law, author of
" A Treatise on the Law of Trusts in India"; and GILBERT S. HEN-
DERSON, M.A., Bar.-at-Law, author of " A Treatise on the Law of Testa-
mentary and Intestate Succession in India." Third Edition, Rs. 14.

THE INDIAN CRIMINAL DIGEST.—Containing all the Important
Criminal Rulings of the various High Courts in India, together with many
English Cases which bear on the Criminal Law as Administered in Indian
In Four Parts: I.—Indian Penal Code. II.—Evidence. III.—Criminal
Procedure. IV.—Special and Local Acts. Vol. II.—1885 to 1893. By
J. T. HUME, Solicitor, High Court, Calcutta, in charge of Government
Prosecutions. Royal 8vo, cloth. Rs. 7-8.

PHILLIPS.—MANUAL OF INDIAN CRIMINAL LAW.—Being the
Penal Code, Criminal Procedure Code, Evidence, Whipping, General
Clauses, Police, Cattle-Trespass, Extradition Acts, with Penal Clauses of
Legal Practitioners' Act, Registration, Arms, Stamp, &c., Acts. Fully
Annotated, and containing all Applicable Rulings of all High Courts
arranged under the Appropriate Sections up to date; also Circular Orders
and Notifications. By H. A. D. PHILLIPS, C.S. Second Edition. Thick
crown 8vo. Rs. 10.

PHILLIPS.—COMPARATIVE CRIMINAL JURISPRUDENCE.—Showing the Law, Procedure, and Case-Law of other Countries, arranged
under the corresponding sections of the Indian Codes. By H. A. D.
PHILLIPS, B.C.S. Vol. I, Crimes and Punishments. Vol. II, Procedure
and Police. Demy 8vo, cloth. Rs. 12.

PRINSEP.—CODE OF CRIMINAL PROCEDURE (ACT X OF 1882), and
other Laws and Rules of Practice relating to Procedure in the Criminal
Courts of British India. With Notes. By the Hon'ble H. T. PRINSEP,
Judge, High Court, Calcutta. Tenth Edition, brought up to June 1892.
Royal 8vo. Rs. 12.

TOYNBEE.—THE VILLAGE CHAUKIDARI MANUAL.—BEING ACT
VI (B. C.) OF 1870, as amended by Acts I (B. C.) of 1871 and 1886. With
Notes, Appendices, &c. By G. TOYNBEE, C.S., Magistrate of Hooghly.
Second Edition, Revised. Crown 8vo, cloth. Re. 1.

THACKER, SPINK AND CO., CALCUTTA.

SWINHOE.—THE CASE NOTED PENAL CODE, AND OTHER ACTS. Act XLV of 1860 as amended with references to all Reported Cases decided under each section. Crown 8vo, cloth.   Rs. 7.

EVIDENCE.

FIELD.—THE LAW OF EVIDENCE IN BRITISH INDIA.—Being a Treatise on the Indian Evidence Act as amended by Act XVIII of 1872. By the Hon'ble C. D. FIELD, M.A., LL.D.   Fifth Edition.   Rs. 18.

STEPHEN.—THE PRINCIPLES OF JUDICIAL EVIDENCE.—An Introduction to the Indian Evidence Act, 1872. By SIR JAMES FITZ-JAMES STEPHEN, formerly Legislative Member of the Supreme Council of India. A New Edition. Crown 8vo, cloth. Rs. 3.

AMEER ALI AND WOODROFFE.—THE LAW OF EVIDENCE APPLICABLE TO BRITISH INDIA.   BY SYED AMEER ALI, M.A., C.I.E., Barrister-at-Law, Judge of the High Court of Judicature and J. G. WOODROFFE, M.A., B.C.L., Barrister-at-Law.   [*In preparation.*]

MEDICAL JURISPRUDENCE.

LYON.—MEDICAL JURISPRUDENCE FOR INDIA.—By J. B. LYON, F.C.S., F.C., Brigade-Surgeon, Professor of Medical Jurisprudence, Grant Medical College, Bombay. The Legal Matter revised by J. D. INVERARITY, Barrister-at-Law. Second Edition. Illustrated. 8vo. Rs. 16.

GRIBBLE.—OUTLINES OF MEDICAL JURISPRUDENCE FOR INDIA. By J. D. B. GRIBBLE, M.C.S. (Retired), PATRICK HEHIR, M.D., F.R.C.S.E., Third Edition, Revised, Enlarged, and Annotated. Demy 8vo. Rs. 5-8.

DIGESTS.

SUTHERLAND.—THE DIGEST OF INDIAN LAW REPORTS.—A Compendium of the Rulings of the High Court of Calcutta from 1862, and of the Privy Council from 1831 to 1876. By D. SUTHERLAND, Barrister-at-Law. Imp 8vo, Rs. 8. Vol. II, 1876 to 1890, thick cloth, imp. 8vo. Rs. 12.

WOODMAN.—A DIGEST OF THE INDIAN LAW REPORTS and of the Reports of the cases heard in Appeal by the Privy Council, 1890 to 1898. Edited by J. V. WOODMAN. Super-royal 8vo. Rs. 16.

WOODMAN.—A DIGEST OF THE INDIAN LAW REPORTS and of the Reports of the cases heard in Appeal by the Privy Council, 1887 to 1889. Edited by J. V. WOODMAN, Barrister-at-Law. Super-royal 8vo, cloth. [*In the Press.*]

THACKER, SPINK AND CO., CALCUTTA.

HINDU AND MAHOMMEDAN LAW.

AMEER ALI.—THE STUDENT'S HAND-BOOK OF MAHOMMEDAN LAW. By the Hon'ble Syed Ameer Ali, c.i.e., author of "The Law relating to Gifts, Trusts, &c." "Personal Law of the Mahommedans," &c., &c. Crown 8vo. Second Edition, Revised. Rs. 3.

AMEER ALI.—MAHOMMEDAN LAW, VOL. I.—By the Hon'ble Syed Ameer Ali, c.i.e., Barrister-at-Law. Containing the Law relating to Gifts, Wakfs, Wills, Pre-emption, and Bailment. With an Introduction on Mahommedan Jurisprudence and Works on Law. (Being the Second Edition of Tagore Law Lectures, 1884. Royal 8vo, cloth. Rs. 16.

AMEER ALI.—MAHOMMEDAN LAW, VOL. II.—By the Hon'ble Syed Ameer Ali, c.i.e., Barrister-at-Law. Containing the Law Relating to Succession and Status, according to the Hanafi, Máliki, Sháfei, Shiah and Mutazela schools, with Explanatory Notes and an Introduction on the Islâmic system of Law. Being a Second Edition of "The Personal Law of the Mohammedans." Revised. Royal 8vo, cloth. Rs. 14.

These two volumes form a complete Digest of the Mahommedan Law.

COWELL.—HINDU LAW.—Being a Treatise on the Law Adminis- tered exclusively to Hindus by the British Courts in India. (Tagore Law Lectures, 1870 and 1871.) By Herbert Cowell, Barrister-at-Law. Royal 8vo, cloth. Lectures, 1870, Rs. 12 ; Lectures, 1871, Rs. 8.

JOLLY.—THE HINDU LAW OF INHERITANCE, PARTITION, AND Adoption according to the Smritis. By Prof. Julius Jolly, of Wurtzburg. (Tagore Law Lectures, 1883.) Royal 8vo. Rs. 10.

THE HINDU LAW OF ENDOWMENTS. Being the Tagore Lectures, 1891. By Pandit Prannath Sarasvati. Royal 8vo cloth.

RUMSEY.—AL SIRAJIYYAH.—Or, The Mahommedan Law of In- heritance, with Notes and Appendix. By Almaric Rumsey. Second Edition, Revised, with Additions. Crown 8vo. Rs. 4-8.

SIROMANI.—A COMMENTARY ON HINDU LAW OF INHERITANCE, Succession, Partition, Adoption, Marriage, Stridhan, and Testamentary Disposition. By Pundit Jogendro Nath Bhattacharjee, m.a., b.l. Second Edition. 8vo. Rs. 16.

WILSON.—INTRODUCTION TO THE STUDY OF ANGLO-MAHOM- medan Law. By Sir Roland Knight Wilson, Bart., m.a., l.m.m., late Reader in Indian Law to the University of Cambridge, author of "Modern English Law." 8vo, cloth. Rs. 6.

WILSON.—A DIGEST OF ANGLO-MUHAMMADAN LAW.—Being an attempt to set forth, in the form of a Code, the rules now actually administered to Muhammadans only by the Civil Courts of British India, with explanatory Notes and full reference to Modern Case-Law, as well as to the ancient authorities. [In the Press.

THACKER, SPINK AND CO., CALCUTTA.

LAW MANUALS, ETC.

COWELL.—THE HISTORY AND CONSTITUTION OF THE COURTS AND LEGISLATIVE AUTHORITIES IN INDIA. Second Edition, Revised. By HERBERT COWELL. 8vo, cloth. [1884] Rs. 6.

HAND-BOOK OF INDIAN LAW.—A POPULAR AND CONCISE STATEMENT OF THE LAW generally in force in British India, designed for non-legal people, on subjects relating to Person and Property. By a Barrister-at-Law and Advocate of the High Court at Calcutta. Crown 8vo pp. xxiv, 754. Cloth gilt. Rs. 12.

CARNEGY.—KACHAHRI TECHNICALITIES.—A GLOSSARY OF TERMS, Rural, Official and General, in daily use in the Courts of Law, and in illustration of the Tenures, Customs, Arts, and Manufactures of Hindustan. By P. CARNEGY. Second Edition. 8vo, cloth. Rs. 9.

CURRIE.—THE INDIAN LAW EXAMINATION MANUAL.—By FENDALL CURRIE, of Lincoln's Inn, Barrister-at-Law. Fourth Edition, Revised. Demy 8vo. [1892] Rs. 5.
  CONTENTS:—Introduction—Hindoo Law—Mahommedan Law—Indian Penal Code—Code of Civil Procedure—Evidence Act—Limitation Act—Succession Act—Contract Act—Registration Act—Stamp and Court-Fees Acts—Mortgage—Code of Criminal Procedure—The Easements Act—The Trust Act—The Transfer of Property Act—The Negotiable Instruments Act.

THE SEA CUSTOMS LAW OF INDIA (ACT VIII OF 1878) with Notes. And the Tariff Act of 1894. By W. H. GRIMLEY, I.C.S., late Secretary to the Board of Revenue, Calcutta. 8vo cloth. Rs. 7-8.

LEGISLATIVE ACTS OF THE GOVERNOR-GENERAL OF INDIA IN COUNCIL OF 1893. With Table of Contents and Index. Royal 8vo, cloth. Rs. 2-8; Previous Volumes available.

DONOGH.—THE STAMP LAW OF BRITISH INDIA.—As constituted by the Indian Stamp Act (I of 1879), Rulings and Circular Orders, Notifications, Resolutions, Rules, and Orders, together with Schedules of all the Stamp Duties chargeable on Instruments in India from the earliest times. Edited, with Notes and complete Index, by WALTER R. DONOGH, M.A., of the Inner Temple, Barrister-at-Law. Demy 8vo, cloth, gilt. With Supplements to 1894. Rs. 8.

GRIMLEY.—AN INCOME-TAX MANUAL.—Being Act II of 1886, With Notes. By W. H. GRIMLEY, B.A., C.S., Commissioner of Income-Tax, Bengal. Royal 8vo. Rs. 3-8. Interleaved, Rs. 4.

ADVOCACY AND EXAMINATION OF WITNESSES.—The work treats of matters of practice such as taking instructions, speech, argument, examination-in-chief and cross-examination, and includes a resumé of the duties and liabilities of Pleaders in India. The Legal Practitioners Act, with the Rules of the High Courts relating to the admission of Pleaders and Mookhtars, appears in the form of an appendix. By H. N. MORISON.

THACKER, SPINK AND CO., CALCUTTA.

INDIAN MEDICAL GAZETTE.

A Record of Medicine, Surgery and Public Health, and of
General Medical Intelligence, Indian and European.

Edited by W. J. SIMPSON, M.D.

Published monthly.   Subscription Rs. 18 per annum.   Single copy Rs. 2.

The *Indian Medical Gazette* was established Twenty-eight years ago,
and has earned for itself a world-wide reputation by its solid con-
tributions to Tropical Medicine and Surgery. It is the Sole representative medium for recording the work and experience of the Medical
Profession in India ; and its very numerous Exchanges with all the
leading Medical Journals in Great Britain and America enable it
not only to diffuse this information broadcast throughout the world,
but also to cull for its Indian readers, from an unusual variety of
sources, all information which has any practical bearing on medical
works in India.

The *Indian Medical Gazette* is indispensable to every member of
the Medical Profession in India who wishes to keep himself abreast
of medical progress, for it brings together and fixes the very special
knowledge which is only to be obtained by long experience and close
observation in India. In this way it constitutes itself a record of
permanent value for reference, and a journal which ought to be in
the library of every medical man in India or connected with that
country. The Transactions of the Calcutta Medical Society, which meets
monthly, is printed *in extenso*, and is a very valuable feature in the
Gazette.

The Gazette covers altogether different ground from *The Lancet*
and *British Medical Journal*, and in no way competes with these for
general information, although it chronicles the most important item
of European Medical Intelligence. The whole aim of the Gazette is
to make itself of special use and value to Medical Officers in India and
to assist and support them in the performance of their difficult duties.

It is specially devoted to the best interests of The Medical Services,
and its long-established reputation and authority enable it to com-
mand serious attention in the advocacy of any desirable reform or sub-
stantial grievance.

The Contributors to the *Indian Medical Gazette* comprise the most
eminent and representative men in the profession.

THACKER, SPINK AND CO., CALCUTTA.

THE JOURNAL OF THE
PHOTOGRAPHIC SOCIETY OF INDIA.
AN ILLUSTRATED MONTHLY JOURNAL.

Invaluable to all lovers of the Art of Photography.

A medium for the earliest information on all discoveries in Photography, Photographic Literature, Experience and News.

The Journal has a large and increasing circulation, is affiliated with Clubs or Amateur Societies all over India, Ceylon, Burma and the Straits Settlements, and has an extensive circulation out of India.

Each number of the Journal is illustrated with a Picture reproduced by a photo-mechanical process.

SUBSCRIPTION—RS. 5 PER ANNUM.

Members of the Society, *free.*

THE RACING CALENDAR.
A FORTNIGHTLY CALENDAR.

Published in accordance with the Rules of Racing, under the authority of the Stewards of the Calcutta Turf Club.

A Record of all Race Performances in India, Racing Fixtures and Racing information, Meetings of the Calcutta Turf Club, Registration of Colours, Assumed Names of Owners, Jockeys' Licences, Unpaid Forfeit List, List of Defaulters, Change in Horses' Names, Horses and Ponies classed, aged and measured, and all information relating to Racing.

ANNUAL SUBSCRIPTION, RS. 12.

THACKER, SPINK AND CO., CALCUTTA.

STANDARD WORKS ON INDIA.

THE JOURNAL OF INDIAN ART.—With Full-page Coloured Illustrations. Folio 15 by 11. Parts 1 to 48 ready. Re. 2 each.

THE SACRED BOOKS OF THE EAST.—Translated by various Oriental Scholars. Edited by F. MAX MULLER. *List of Volumes on application.*

THE FAUNA OF BRITISH INDIA.—Including Ceylon and Burma. Published under the authority of the Secretary of State for India. Edited by W. T. BLANFORD, F.H.S., and Illustrated.

Mammalia.	By W. T. BLANFORD, F.H.S.	...	20s.
Fishes, 2 vols.	By Dr. FRANCIS DAY	...	40s.
Birds, vols. I and II.	By F. W. OATES	...	35s.
Reptilia and Batrachia.	By G. A. BOULENGER	...	20s.
Moths, 2 vols.	By F. HAMPSON	...	40s.

THE INDIAN MUTINY, 1857-58.—SELECTIONS FROM THE LETTERS, DESPATCHES, and other State Papers preserved in the Military Department of the Government of India. Edited by GEORGE W. FORREST, B.A., Director of Records of the Government of India. With a Map and Plans. Vol. 1. Delhi. Royal 8vo. Rs. 10.

WARREN HASTINGS.—SELECTION FROM THE LETTERS, DESPATCHES, and other State Papers preserved in the Foreign Department of the Government of India, 1772-1785. Edited by GEORGE W. FORREST, B.A. 3 vols. Fcap., cloth. Rs. 12.

THE ADMINISTRATION OF WARREN HASTINGS, 1772-1785.—Reviewed and Illustrated from Original Documents. By G. W. FORREST, B.A. 8vo, cloth. Rs. 4.

ANNALS OF RURAL BENGAL.—By W. W. HUNTER, C.I.E., LL.D. 8vo. Rs. 6.

ILLUSTRATION OF SOME OF THE GRASSES OF THE SOUTHERN PUNJAB.—Being Photo-Lithograph Illustrations of some of the principal Grasses found at Hissar. With short descriptive letter-press. By WILLIAM COLDSTREAM, B.A., B.C.S. Illustrated with 89 Plates. Demy folio. Rs. 16.

ILLUSTRATIONS OF INDIAN FIELD SPORTS.—Selected and Reproduced from the Coloured Engravings first published in 1807 after designs by Captain THOMAS WILLIAMSON, Bengal Army. Small oblong, handsome cloth cover. Printed in colours. Rs. 9.

THACKER, SPINK AND CO., CALCUTTA.

Demy 8vo. 400 pages, Rs. 14.
HANDBOOK OF THE
TIBETAN COLLOQUIAL LANGUAGE.
In Three Parts.

I.—Grammar of the Colloquial.
II.—Conversational Exercises with Technical, Mythological, Zoological and Geographical Lists.
III.—Compendious Vocabulary, Ladaki—Central Tibetan—Literary Tibetan.

By Rxv. GRAHAM SANDBERG, *Chaplain, H. M.'s Indian Service.*

Crown 8vo, half morocco, gilt top. 14s.
CONSTABLE'S HAND ATLAS OF INDIA:
A new Series of Sixty Maps and Plans prepared from Ordnance and other Surveys under the direction of J. G. BARTHOLOMEW, F.R.G.S.

Alphabetical Index to Maps and Plans:

Aden.
Afghan Frontier.
Afghanistan.
Agra.
Ajmere.
Allahabad.
Andamans.
Animal Products.
Arakan.
Assam.
Baluchistan.
Bengal.
Berar.
Bhutan.
Bokhara.
Bombay.
Bombay, Environs.
Bombay, Plan of.
British Baluchistan.
Burma, Lower.
Burma, Upper.
Calcutta, Environs.
Calcutta, Plan of.
Canals.
Cawnpore, Plan of.
Central India.
Central Provinces.
Ceylon.
China.
Chinese Turkestan.
Darjeeling.
Delhi.
Farther India.

Geology.
Haidarabad.
India. Key to Sectional Maps
Indian Ocean.
Irawadi.
Johore.
Kafiristan.
Karachi.
Karenni.
Kashmir.
Lahore.
Landaur.
Land Service Elevation.
Land Service Features.
Language Map.
Lucknow.
Lushai Hills.
Madras.
Madras, Plan of.
Malacca.
Manipur.
Matheran.
Military Map.
Mineral Products.
Mission Stations.
Murree.
Mussooree.
Mysore.
Naini Tal.
Nepal.
Nicobars.
North-West Provinces.

Ootacamund.
Orissa.
Oudh.
Pahary.
Pegu.
Penang.
Perak.
Political Map.
Population Map.
Province Wellesley.
Punjab.
Punjab, Eastern.
Races, Prevailing.
Railway Map.
Railways.
Rainfall.
Rajputana.
Rangoon.
Routes to India.
Selangar.
Siam.
Sikkim.
Simla.
Sind.
Socotra.
Srinagur.
Straits Settlements.
Telegraphs.
Temperature.
Tenasserim.
Tibet.
Travancore.
Trichinopoli.

CONSTABLE'S ORIENTAL MISCELLANY.

BERNIER'S TRAVELS IN THE MOGUL EMPIRE.—An entirely new edition, with a Frontispiece printed in 18 colours on Japanese paper, other Illustrations, and Three Maps. By ARCHIBALD CONSTABLE, Mem. As. Soc., Bengal, F.S.A. Scot. Crown 8vo. pp. liv, 500. Rs. 6.

POPULAR READINGS IN SCIENCE.—By JOHN GALL, M.A., LL.B., late Professor of Mathematics and Physics, Canning College, Lucknow, and DAVID ROBERTSON, M.A., LL.B., B.SC. With 56 Diagrams, a Glossary of Technical Terms, and an Index. Crown 8vo. pp. 468. Rs. 5.

AURENG-ZEBE : A TRAGEDY.—By JOHN DRYDEN ; and Book II of THE CHACE : a Poem by WILLIAM SOMERVILE. Edited, with Biographical Memoirs and Copious Notes, by K. DEIGHTON, B.A., editor of 'Select Plays of Shakespeare.' With a Portrait of Dryden, and a Coloured Reproduction of an Indian Painting of the Emperor Akbar Deer-stalking. Crown 8vo. pp. xiii, 222. Rs. 5.

LETTERS FROM A MAHRATTA CAMP.—By THOS. D. BROUGHTON. A new edition, with an Introduction by the Right Hon Sir M. E. GRANT DUFF, G.C.S.I., F.R.S. Notes, Coloured and other Illustrations, very full Index, and a Map. Rs. 6.

RAMBLES AND RECOLLECTIONS OF AN INDIAN OFFICIAL.—By Major-General SIR W. H. SLEEMAN, K.C.B. A new edition. Edited by VINCENT ARTHUR SMITH. I.C.S. With a copious Bibliography, Index and a Map. 2 vols. Rs. 12-12.

STUDIES IN MOHAMMEDANISM.—Historical and Doctrinal, with a Chapter on Islam in England. By JOHN J. POOL. With a frontispiece and Index. Crown 8vo, cloth. Rs. 6.

"As a 'popular text-book,' dealing with some of the most picturesque aspects of Islam, it deserves more than ordinary attention."—*Times.*

THE GOLDEN BOOK OF INDIA.—A Biographical and Statistical Directory of the Ruling Princes, Chiefs, Nobles, and Titled and Decorated Personages of the Indian Empire; the dates of their birth and succession to the *guddi*; a concise account of their immediate predecessors and all the deeds of honour and valour of their house and family. Imp., red cloth, gilt. Rs. 35.

THACKER, SPINK AND CO., CALCUTTA.